CLIMATE CHANGE AND DISEASES

CLIMATE CHANGE AND DISEASES

Edited by

Dr. Pralhad Jaybhaye

Delve
PUBLISHING

www.delvepublishing.com

Climate Change and Diseases

Dr. Pralhad Jaybhaye

Delve Publishing
4164 Lakeshore Road
Burlington, ON L7L 1A4
Canada
www.delvepublishing.com

Email: orders@arclereducation.com

ISBN: 978-1-77956-357-6 (Hardcover)

ABOUT THE EDITOR

Dr. Pralhad Jaybhaye (1972) is presently serving as Associate Professor, Department of Agriculture Meteorology, Vasantrao Naik Marathwada Krushi Vidyapeeth, Parhani, (Maharashtra). He did his B. Sc.(Agri.) in 1995 and M. Sc. (Agriculture Meteorology) 1998 from Mahatma Phule Krushi Vidyapeeth, Rahuri (College of Agriculture, Pune) and Ph. D. (Agril. Meteorology & Physics) in 2020 from BCKV, Mohanpur, West Bengal. He is recognized a Post Graduate Teacher and Research Guide for M. Sc. And Ph. D. (Agriculture Meteorology).

He started his career as a Technical Officer (Agrometeorology), NARP, Igatpuri (MPKV, Rahuri) from 2000 to 2008. His field of specialization is Agrometeorology and his area of research interest are Micrometeorology, Agricultural Water Management, Climate Change and Extension Agrometeorology. In addition to this, he has performed additional duties as a Nodal officer, Commercial Fruit Nursery, PDKV, Akola (2005-2008); Principal Nodal officer, GKMS & FASAL project (2008-2016) and Principal, Agriculture Technology School, Jalna (VNMKV, Parbhani) (2020-2023). He had conducted National workshop, farmer training programs, workshops and many farmer's rallies/awareness progrommes. During his total 25 year's tenure of services, under his guidance 27 students completed M.Sc. (Agrometeorology) and he has co-guided for more than 25 students to the same degree course. He has published one Text Book (Handbook of Agrometeorology); 2 Subject Books (1-Hydrometeorology, 2-Weather and Climate Monitoring and Prediction Systems in Context of Agriculture) and other 6 books related to Agrometeorology; 25 research papers & 05 short communications in international and national reputed journals.

He has participated and presented research papers more than 25 International, National, State Seminar, Workshops and Symposiums. Besides these he had done extension Agricultural Meteorology activities like TV talk (17) & Radio talks (200), guest lectures in scientist training & farmer training programs had delivered. Dr. Pralhad Jaybhaye has also written more than 450 popular articles in reputed regional language periodicals and daily newspapers.

He had donated Gold Medal to M. Sc. (Agriculture Meteorology), VNMKV, Parbhani in the memory of his father (Late Freedom Fighter Mr. Rambhaji Temaji Jaybhaye) for enrichment capacity building as well as raise academic interest in the Agriculture Meteorology area.

TABLE OF CONTENTS

LIST OF FIGURES

PREFACE

The Earth's climate system is influenced by several natural variables. Greenhouse gases in the atmosphere have a heating effect on the surface of the Earth. This warming effect has been recognized by scientists for about 200 years. These gases redirect the heat energy that is released back toward the surface by absorbing it. Without this greenhouse effect, Earth would be a freezing and probably dead planet. Earth's orbit around the sun is another natural element influencing temperature, with timescales estimated in millennia. These changes cause the surface to receive different amounts of solar radiation. Minute differences in the brightness of the sun or radiation cause slight variations in the surface temperature. A probable adjustment could be modifying our farming methods to align with the evolving local patterns of precipitation and temperature. Building coastal defenses against the flooding caused by sea level rise would be another. The most hazardous alterations, however, can only be avoided by mitigation. Geoengineering is one mitigating strategy that has received a lot of attention lately. By manipulating the climate system, the heating effect of increasing greenhouse gas concentrations can be partially or completely mitigated. It is commonly known that the environment has been changing over the past few decades. In a nutshell, they are brought on by large-scale atmospheric emissions of carbon dioxide, methane, and nitrous oxide, which are partly caused by the burning of fossil fuels to harvest electricity and the large-scale production of domestic animals for human use or consumption. An increase in global temperature and a decrease in dissolved oxygen in water are some consequences of climate change. The metabolism, development, growth, reproduction, and lifespan of aquatic species, including fish, are impacted by these abiotic changes.

This book introduces climate change with focus on diseases. This book can help all the students who are interested in pursuing research in climate change fields.

INTRODUCTION

We are primarily concerned with natural climatic changes and their consequences for the ecosystem and society during the last 10,000 years, even though global warming is not disregarded as an anthropogenic climate change. The surprising scientific advancements in the last ten years in the fields of genetics, modeling, climate sciences, and archaeological evaluation served as the impetus for this effort. It is astounding to learn that comprehensive historical environmental data is preserved in natural archives that can be accessed and scrutinized today. This calls for a stronger synergy and resilience between the natural sciences and humanities. Climate change is unquestionably a serious problem and a hot topic in today's world. Their impacts extend beyond human health to include changes in the morphology, phenology, reproduction, health, and general productivity of the farming system and plants. It devastates the ecosystem structure and functioning. Chlorofluorocarbons (CFCs), nitrous oxide, carbon dioxide, and methane are known to be strong greenhouse gases (GHGs) that contribute to global warming and impact the economy and productivity of agro-ecosystems. The negative effect is the loss of biodiversity that results from events like plant pathogen outbreaks, changes in land usage, and modifications to the structure and composition of vegetation. These issues are extremely problematic for the functioning of the entire ecosystem. Small temperature variations can have remarkable effects on the systems that support human livelihoods, like crop output and water availability. Improved crop management techniques combined with other components could effectively slow down and prevent global climate change. Furthermore, addressing the issues of global warming and present food security can be achieved by reforming agricultural sectors.

Chapter 1

INTRODUCTION TO CLIMATE CHANGE

Contents

1.1 INTRODUCTION

The annual rise rate of carbon dioxide (CO_2) concentration has been at its greatest. The amount of CO_2 in the atmosphere today is far higher than it has been throughout the previous 650,000 years(Galbraith & Eggleston, 2017). CO_2 makes up around 77 percent of all greenhouse gases produced by humans, making it the most significant. The Earth's climate has clearly changed as a result of the concentrations of the three main greenhouse gases, CO_2, methane, and nitrous oxide, which are at their highest points in at least 10,000 years(Uprety et al., 2019). Eleven of the past twelve years (1995–2006) rank among the twelve hottest years since 1850(Brunet et al., 2007). The global surface temperature has risen by 0.74°C over the last 100 years (1906–2005), increasing more quickly over land than oceans(Prabhjyot-Kaur & Hundal, 2010). Naturally, there will be variations in the climate. For example, in February 2008, the National Oceanic and Atmospheric Administration (NOAA) predicted that La Niña conditions, which are the cold phase of the El Niño Southern Oscillation (ENSO), would persist throughout spring 2008.

However, by May, La Niña started to shift towards more ENSO-neutral conditions. But the long-term warming trend continued after taking into consideration these documented climatic variations. The combination of the oceanic thermal expansion and the melting of more glaciers and polar ice sheets is what causes sea levels to rise. Precipitation decreases in the Sahel, Mediterranean, southern Africa, and portions of southern Asia between 1900 and 2005 are among the other prominent alterations in the climate. Since the 1970s, majority of the regions have experienced longer and more severe droughts, especially in the tropics and subtropics. The IPCC concludes that variations in drought have been exacerbated by drought brought on by rising temperatures and falling precipitation. In contrast, eastern regions of North and South America, as well as northern Europe and northern and central Asia, have experienced a rise in precipitation. Since 1970, there has seemed to be an increase in the North Atlantic cyclone activity.

The IPCC has projected warming of roughly 0.2°C each decade for the next two decades under a variety of emission scenarios(MacIsaac, 2019). Following that, the paths taken by greenhouse gas emissions under various scenarios of socioeconomic development and the application of mitigation techniques varies strikingly. It has been known that differences in the climate, especially in terms of temperature, precipitation, and extreme weather, can have an impact on health. Interest in the connections between climate and health has increased as a result of growing scientific consensus regarding the certainty of global climate change. Many methodologies are available for investigating the possible health effects of climate change. These include research on the relationships between historical climate variability and disease, the relationships between recent trends in climatic variables and disease epidemiology, and the reactions of vector species to variations in rainfall and temperature. Furthermore, a variety of methods have been developed to estimate the possible long-term effects of climate change on health. The necessity to take confounding variables into account as potential explanations for apparent connections between climatic variables and health outcomes is one of the methodological issues in the research of potential relationships between climate change and health. Confounding variables could be anything from shifts in insecticide resistance (vector-borne diseases) to shifts in drug resistance (malaria); population migration, which could expose non-immune populations to infectious diseases; or shifts in the effectiveness of disease surveillance systems over time. Modeling the anticipated impact of climate change is made more difficult by the changes in local transmission patterns that some diseases, like malaria, exhibit. Future links between the changing environment and health outcomes may be lessened by advancements in public health infrastructure that enhance adaptability to climate change.

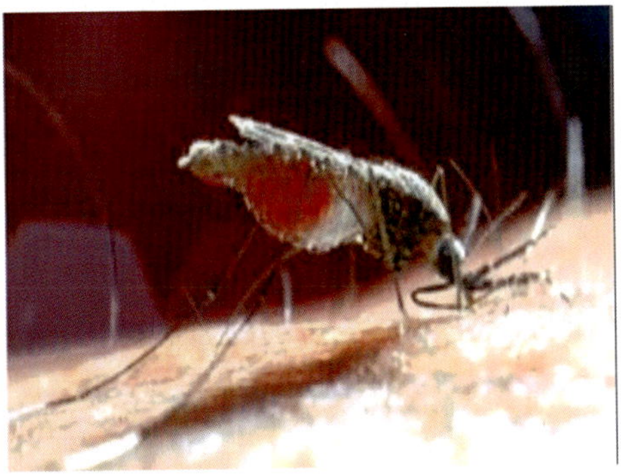

***Figure 1.1** Anopheles darlingi*

Source: https://en.wikipedia.org/wiki/Anopheles_darlingi#/media/File:Bioswiki.jpg, *Creative Commons License.*

Although the majority of our knowledge on the connections between climate and health is derived from studies of short-term variability, climate change is most likely a long-term process that will unfold over decades or perhaps centuries. There is a good chance that additional environmental changes, like deforestation, greater mobility of the local population, increasing international travel, and depletion of aquifers in some areas, will interact with climate change. Deforestation, for instance, has the potential to alter the distribution of disease vectors and exacerbate climate change. Additionally, population movement into formerly forested areas may heighten exposure to many diseases. In Peru, deforested areas have a hundred-fold higher abundance of the malaria vector *Anopheles darlingi*(Figure 1.1) than untouched/pristine forest areas, and this difference in abundance cannot be primarily attributable to increased population density(Bharti et al., 2006). Deforestation may lower the risk of malaria in Southeast Asia while increasing it in the Americas and Africa.

1.2 WATER

Water-related diseases include those that are carried by the water due to pathogens ingested from contaminated water and those that are washed down by the water due to inadequate hygiene. A global overview of cross-sectional studies based on published reports from low- and middle-income countries between 1954 and 2000 indicates an average 4% increase in the incidence of diarrhea in children under the age of five for every 10 mm decrease in monthly rainfall(Fischer Walker et al., 2012). Since over 2 billion people now reside in largely arid areas of the earth, their lack of access to clean water is probably going to affect them excessively.

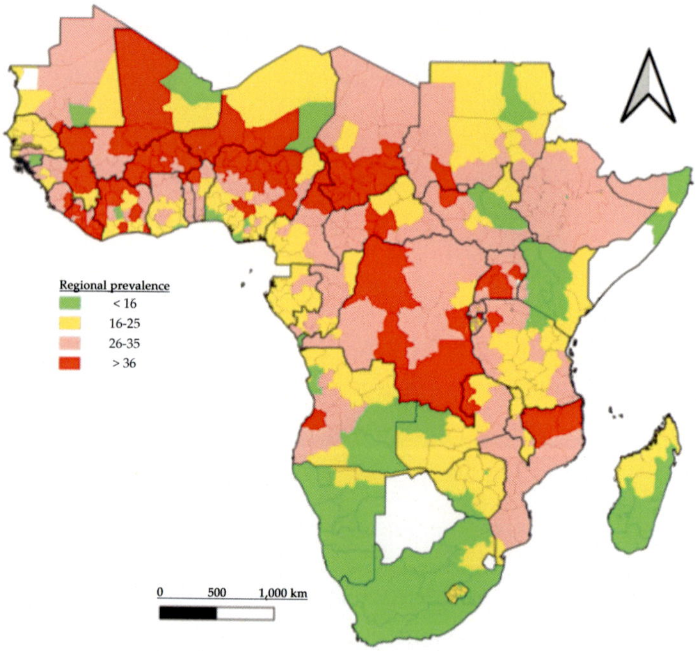

Figure 1.2 Child mortality

Source: Hailu, B.A., Ketema, G. & Beyene, J. Mapping of mothers' suffering and child mortality in Sub-Saharan Africa. Sci Rep 11, 19544 (2021). https://doi.org/10.1038/s41598-021-98671-9, *Creative Commons License.*

Child mortality (Figure 1.2) is still too high, especially in sub-Saharan Africa, despite significant advancements in the treatment of diarrheal disease, mainly due to the extensive use of oral rehydration therapy. Nearly 2 million children under the age of five die from diarrheal ailments each year. It's crucial to remember that the majority of freshwater is utilized for irrigation rather than human consumption; Therefore, there may be no clear link between decreased freshwater availability and diarrhea. Hand washing with soap protects against diarrhea, and in areas with low rainfall, less hand washing may occur due to decreased freshwater availability and/or higher expenses. Epidemics of water-borne diseases such as cryptosporidiosis have been linked to heavy precipitation. While the aforementioned global overview did not find any correlation with temperature, other research has linked higher temperatures to an increase in episodes of diarrheal ailments in Australia, Peru, and the Pacific Islands. The Bay of Bengal has provided the strongest evidence for the link between sea surface temperatures and cholera transmission. A time-series analysis of weekly salmonellosis cases using data from sixteen industrialized sites revealed a roughly linear increase in cases reported over the 6°C threshold(Hazards (BIOHAZ), 2010). With a week or so gap between the temperature increase and the rise in cases, the link seemed to be especially noticeable in the case of *Salmonella enteritidis*, indicating that the influence may be on the replication of salmonella after food has been prepared. The link between Meningococcal meningitis epidemics in West Africa and the arid northern winds known as the Harmattan is among the best-documented instances of how climate can influence the start of disease. The disease may be caused by mucosal abrasion and dryness due to strong winds carrying dust, while the exact cause of the condition is unknown.

There is evidence that the geographic range of Meningococcal meningitis in West Africa has widened recently, potentially due to changes in climate and land use. Large-scale ocean-atmospheric climatic phenomena known as El Niño episodes have been happening for thousands of years. El Nino cycles, which typically last between three and seven years, are linked to climate shifts in Australia, Indonesia, the Pacific highlands, East Africa, and some regions of Latin and North America. Often, a La Nina event occurs after an El Nino event. The balance of air pressure variations between the east and west Pacific is called the Southern Oscillation, which is linked to the El Nino phenomena and gives rise to the full name El Nino/Southern Oscillation (ENSO). The primary climate cycle responsible for natural disasters is ENSO. Based on data analysis over three decades between 1963 and 1993, an average El Nino year affects about 35 individuals per 1,000 worldwide, more than four times that of non-El Nino years(Kovats et al., 2003). In many regions, there is a substantial correlation between El Nino and heavy rainfall; but, unlike drought, this correlation is not observed globally since flood-related disasters are comparatively limited, and the risk is elevated during both El Nino and La Nina phases in various parts of the world. Typhoons are more likely to occur near the Marshall Islands in the Pacific Ocean during an El Nino event compared to other times due to the western shift in storm tracks. El Nino events also cause storm activity in parts of the Pacific to increase and decrease in the Atlantic, resulting in hurricanes in the Caribbean and the Gulf of Mexico being less common during the event but more common during a La Nina.

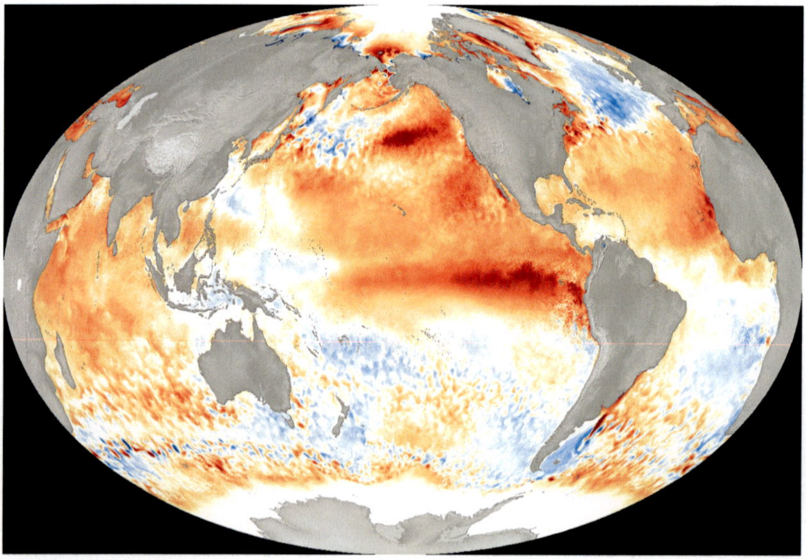

Figure 1.3 ENSO

Source: https://psl.noaa.gov/enso/img/nesdis-el-nino.png, *Creative Commons License.*

1.3 EL NINO/SOUTHERN OSCILLATION

Understanding the possible effects of climate change can be enhanced by examining the association between ENSO (Figure 1.3) events and health outcomes. Research offers valuable evidence for causal relationships between climatic variability and human health. The exact effect of climate change on the El Nino phenomenon is unknown, but future occurrences of such events will take place against a backdrop of climate change, which is anticipated to have more negative effects on society and health. To deduce a causal association between the ENSO cycle and health, relevant climate data for the study location must be obtained. Additionally, there must be a tenable biological relationship between the health outcome in question and climatic variability. Appropriate statistical analysis is needed, accounting for autocorrelation and utilizing long-term datasets spanning decadal periods when possible. Numerous case reports have also linked individual ENSO episodes to disease epidemics; nevertheless, the evidence of this association is not as strong as it may be. The complexity of analyzing the relationships between health and El Nino occurrences stems from their differences in intensity and regional implications. For instance, the southern African region did not experience the predicted drought in 1997 and 1998, and several regions saw above-average rainfall in those years(Reason et al., 2005). Although the data available is limited, several strong correlations have been reported globally, especially in relation to epidemic malaria in nations like Guyana, Colombia, and Venezuela. El Nino-related increased temperatures may promote the spread of malaria in regions like northern Pakistan. Increased rainfall in the post-El Nino and La Nina years has historically raised the risk of epidemic in many desert locations, including the Thar Desert and the Punjab. On the other hand, droughts linked to El Nino have been linked to malaria outbreaks in Venezuela, Colombia, and Sri Lanka. This may be because of decreased river flows that allowed mosquitoes to spawn more frequently. El Nino years in Indonesia have also been linked to the annual occurrence of dengue. On certain Pacific Islands, but not all of them, monthly dengue cases have been linked to the Southern Oscillation index. A correlation was shown between the Southern Oscillation index and the years of the Ross River virus epidemic in Australia from 1968 to 1998, spanning 70 years(Gage et al., 2008). However, some investigations found no correlation between the index and the annual occurrence of the disease. Localized relationships between climatic conditions and health outcomes might not be discernible from aggregated data.

In Dhaka, Bangladesh, cholera and ENSO have been linked, and in recent decades, this connection has been more evident. There was a substantial correlation between the annual incidence of visceral leishmaniasis in Bahia State, northeastern Brazil, and the sea surface temperature in the Pacific Ocean. This correlation was most likely caused by the ENSO cycle and northeastern drought. In 1993, the southern United States experienced an increase in rainfall due to the El Nino event. This was followed by a drought that made it easier for hantavirus pulmonary disease to spread. In addition, the number of cases reported linked to the El Nino that occurred in 1997–1998 increased fivefold. A consistent relationship between ENSO and rodent population increases, which would raise the risk of hantavirus transmission, has not been proven, despite a possible explanation for how climatic conditions can cause significant increases in rodent population. Similar to this, there is strong evidence that large rainfall events in the arid grasslands of East Africa cause outbreaks of Rift Valley fever; yet, there is no correlation between this disease in Kenya and the ENSO index. According to case reports, there may be a correlation between infectious disease outbreaks and extreme weather events like floods, droughts, and storms. However, this correlation is not always evident and is probably dependent on socioeconomic variables, infrastructure damage, and the degree of population displacement. The likelihood of flooding is anticipated to rise due to several mechanisms brought about by climate change. Sea level rise, an increase in the frequency of heavy precipitation episodes, and glacier melting are a few of these. Nonetheless, a thorough analysis of the consequences of flooding in Europe revealed that outbreaks of infectious diseases were rarely a significant threat to public health. A greater prevalence of common mental disorders (depression and anxiety) after flood exposure appears to be a more consistent finding. This is most likely the result of growing poverty, forced relocation, losing source of income, and losing familiar possessions. However, both high-income and low-income countries have documented increases in respiratory and diarrheal ailments after floods. Population displacement and crowding may facilitate the spread of the disease. Leptospirosis, hepatitis, and vector-borne diseases like malaria and Rift Valley fever can all have epidemics, especially in low-income nations. Sometimes floods had the reverse effect and washed away vector breeding places, as was the case in Tanzania in 1997–1998.

Figure 1.4Hurricane Katrina

Source: https://en.wikipedia.org/wiki/File:Katrina_2005-08-28_1700Z.jpg, *Creative Commons License.*

1.4 CLIMATE EFFECTS

Low-lying (areas) populations are especially susceptible to the consequences of sea level rise in low-income nations where flood barriers may not have been updated. For instance, if there was a 4°C temperature increase, 100 cm of sea level rise, an increase in monsoon precipitation, and increased discharge into major rivers, up to 57% of people living in unprotected dry land areas of Bangladesh could be submerged(Penning-Rowsell et al., 2013). Seawater intrusion may cause freshwater aquifers to become salinized. As evidenced by Hurricane Katrina (Figure 1.4) effects on New Orleans, heavily populated urban areas are vulnerable to sea level rise even in developed countries. Due to fecal pollution of drinking water, the incidence increased after Hurricane Katrina. Drought in developing countries may have a wide range of possible health implications. A flow of water adequate to meet the domestic water needs of about 1,000 people would only be enough to irrigate 1 hectare of land, which is enough to feed a few families. It is important to remember that domestic water consumption accounts for only 2% of the world's total water supply. It has also been noted that, when it comes to determining health benefits, research from low-income countries indicates that easy access to water, which leads to increased amounts used for hygiene, is probably more significant than water quality improvements alone. This is probably because most endemic diseases spread from person to person through contaminated hands, food, and other objects due to inadequate hygiene. Study conducted in sub-Saharan African nations revealed that HIV/AIDS may intensify the impact of drought on undernutrition in children, especially in periurban and urban populations with high HIV prevalence. Drought-related malnutrition makes people more vulnerable to diseases like measles, especially in countries with low immunization rates like Somalia.

There are differing impacts of drought on the incidence and distribution of vector-borne diseases. For instance, a decrease in mosquito activity during a drought may be followed by an increase in disease transmission after the drought ends due to the greater number of susceptible hosts. In other situations, water that has stagnated in pools may temporarily enhance the spread of malaria. Extended periods of drought could lead to the reduction of regions that are conducive to the spread of malaria. The likely impact of climate change on malnutrition is difficult to assess because of the variety of factors that affect food production and consumption. These factors include agricultural practices, the potential benefit of carbon dioxide fertilization in raising crop yields, and the ability of disadvantaged populations to purchase food. Although this is not a certain outcome and will depend on the policy decisions that are taken, concerns have also been raised over the possible competition for land between biofuels and agricultural production for food. The areas currently most susceptible to food insecurity, especially sub-Saharan Africa, are expected to be most vulnerable to increased malnutrition as a result of climate change. Natural disasters cause more deaths and direct economic losses in terms of national per capita income in low-income countries than in high-income countries. For instance, research on the effects of Hurricane Mitch (Figure 1.5) on the rural poor in Honduras revealed that one in two of the households had to pay for housing, medical care, or other expenses as a result of the hurricane. Less than ten percent of the losses suffered by households were made up of relief. These financial losses exacerbate poverty and have a significant negative impact on public health as a result of climate-related calamities. A 2000 estimate of Mozambique flood-related mortality revealed, for instance, that a significant portion of the total flood-related deaths were caused by an increase in infant mortality that was linked to a roughly 14% decline in the provinces that experienced flooding(Alderman et al., 2012).

Figure 1.5 Hurricane Mitch

Source: https://en.wikipedia.org/wiki/Hurricane_Mitch#/media/File:Mitch_1998-10-26_1845Z.png, *Creative Commons License.*

1.5 DISEASE SPREAD

There is strong evidence that the Northern Hemisphere spring pollen season is starting earlier due to climate change, which has affected the seasonality of allergic rhinitis. Climate change is considered to have at least increased the chance of a heat wave like the one that occurred in 2003. The summer of 2003 was likely the hottest in Europe since 1500. Over infectious diseases, there is still disagreement over the extent to which changes in the frequency and spread of disease have been caused by climate change. This can be attributed to several additional reasons, including shifting land use patterns, changes in human behavior, and methodological problems with the use and interpretation of relevant climatic data. Sweden has seen shifts in the distribution of ticks toward the north, and since the mid-1980s, the country has seen a significant rise in the incidence of tick-borne encephalitis (TBE) (Figure 1.6)(Holmgren & Forsgren, 1990).

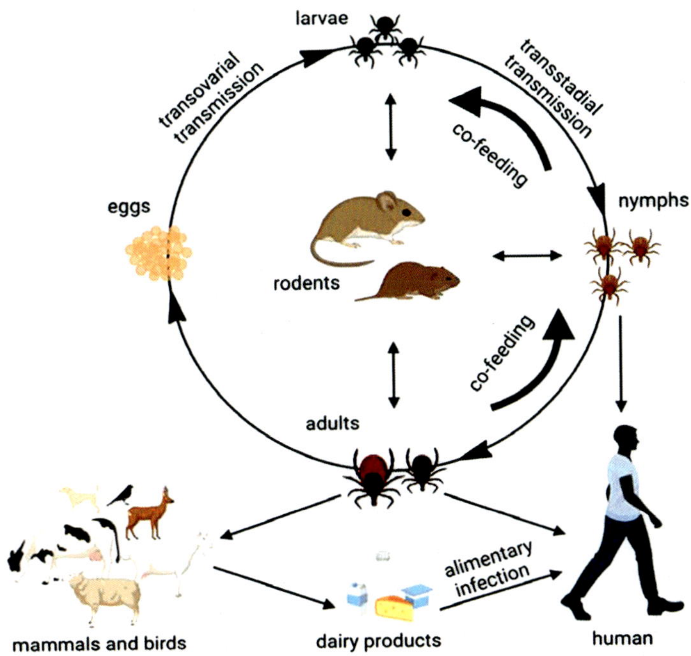

Figure 1.6 Tick-borne encephalitis

Source: Ličková M, Fumačová Havlíková S, Slávikova M, Klempa B. Alimentary Infections by Tick-Borne Encephalitis Virus. Viruses. 2022; 14(1):56. https://doi.org/10.3390/v14010056, *Creative Commons License.*

According to studies conducted on TBE cases in Stockholm County between 1960 and 1998, milder winters and an earlier start to spring may have contributed to the higher incidence of TBE, but other factors, such as changes in land cover and use that increase the number of wildlife hosts for ticks and the number of people living in endemic areas, may also be involved. The possibility that climate change contributes to the rise in malaria cases in the highlands of East Africa in recent decades has attracted a lot of attention. Researchers claimed that there were no significant trends in climate variables and, as a result, concluded that climate change had not contributed to the recurrence of malaria in the region. It was claimed that the apparent lack of a correlation could not be construed as strong proof that climate change had not played a role, and the use of low-resolution data to examine the relationship was challenged. There was evidence of substantial warming trends of about 0.5°C at all areas using both nonparametric and parametric statistical techniques. Climate effects contribute to malaria outbreaks in the highlands of East Africa. Significant spatial variation in the sensitivity of malaria cases to climate variability was found in highland sites in East Africa where seasonality, climate variability, and the number of monthly cases of the disease were studied over decades. The increased weight of climatic disasters, more than population growth alone can account for, has been reported in the most recent Human Development Report published by the United Nations Development Programme (UNDP). The majority of people in poor nations lack insurance, therefore these losses are primarily felt there. Weather-related insurance losses are rising more quickly than population and inflation. In contrast to less than 50 million annually between 1975 and 1979, more than 250 million people were impacted by hydrometeorological disasters, including increases in floods, droughts, lightning strikes, and the intensity of tropical cyclones between 2000 and 2004. Increased sea surface temperatures have been linked to an increase in Atlantic hurricane activity in recent years. The IPCC has evaluated an increasing number of studies that have predicted the anticipated effects of climate change on health.

Under the World Health Organization (WHO), efforts have been made to evaluate the global burden of disease resulting from climate change. The influence of climate change on various health outcomes like malaria, diarrhea, malnutrition, flood mortality, and direct effects from heat and cold was estimated using epidemiological models. There may be some positives, such as a decrease in cold-related mortality in temperate zones, but they will be considerably outweighed by the detrimental effects on health, especially concerning infectious diseases and malnutrition in low-income countries. The negative consequences of climate change may be mitigated by rising income and some degree of adaptation, but in the long run, it is expected that the burden of disease associated with the changes in climate will rise significantly and will be concentrated in the poor populations. However, populations worldwide are likely to face some negative effects, especially if temperature increases surpass 2°C, at which point the likelihood of major negative events, like ice caps melting and ecosystem disruption appears to be unacceptably high. Physiological and behavioral adaptation to climate change may occur without changes to policy, but adaptation strategies can improve population resilience to future climate change and have a positive impact on public health. In order to increase the likelihood of achieving the UN Millennium Development Goals, policies that enhance access to clean water and sanitation, encourage hygiene practices, encourage vaccination uptake, and strengthen health systems are required. The burden of malaria can be decreased by increasing the use of insecticide-impregnated nets, choosing antimalarial medication combinations that account for current drug resistance patterns, and implementing efficient vector control techniques like indoor spraying.

1.6 HEATWAVES

Interest in climate-based early warning health systems for heat waves and climate-sensitive diseases has increased as a result of the threat posed by climate change. If early warning systems are to be effective, local health systems must incorporate them. The majority of standard disease surveillance systems are unable to reliably and promptly report spikes in the number of malaria cases. Because disorders other than malaria must be taken into consideration as potential causes of outbreaks, there is a need to enhance the routine data collection on parasitological confirmed cases of malaria. Seasonal predictions have the potential to enhance readiness for climatic fluctuations and severe incidents linked to phenomena like El Nino; these methods were employed to alert certain governments during the development of a powerful El Nino in 1997–1998. Approximately 1.6 billion people lived without access to electricity in 2006, while 2.4 billion families rely on solid fuels like coal, wood, and manure(Kaygusuz, 2011). In addition to reducing exposure to high levels of indoor air pollution, which has an estimated 1.6 million annual deaths linked to it, meeting the basic energy needs of the poor will need around 1% of the current energy use. A coordinated effort is required to increase the availability of more energy-efficient stoves and help poor communities switch to cleaner fuels like biogas, liquefied petroleum gas, or kerosene. Utilizing renewable energy sources whenever feasible, electrification can enhance adaptation by delivering electricity to keep vaccines cold chained, to give medical facilities a dependable power source, and to enable the use of ICTs (information and communication technologies). An important concern for the modern era is the requirement for policies that serve the energy demands of the poor while preventing dangerous anthropogenic interference with the climate.

The majority of renewable energy technologies can improve health in the short term by lowering air pollution exposure and reducing greenhouse gas emissions. For instance, there is evidence of several public health problems associated with dams, which can be utilized to produce hydroelectricity and to support climate change adaptation through better irrigation. These issues include rising rates of malaria and schistosomiasis as well as the introduction of Rift Valley fever. Future water resource development projects should involve a thorough assessment of potential effects, even though it was not possible to estimate the attributable fraction of the malaria burden owing to dam building and irrigation. Leptospirosis, cholera, dengue fever, and malaria became more common. A six-week period of rain and three storms in 2000 submerged Mozambique, resulting in a five-fold increase in malaria cases(Sorensen et al., 2015). Tens of thousands of people lost their lives in 2003 as a result of a summer heat wave that devastated Europe, melted 10% of the Alpine glacial mass, withered crops and started forest fires. More than 200 American cities saw temperatures surpass previous records. The poor people in Phoenix, Arizona, suffered greatly as a result of 39 days with temperatures above 100°F (38°C), including one week with temperatures above 110°F (43°C)(Harlan et al., 2014). Extreme weather incidences of this kind are a reflection of the significant and continuous changes in our climate, to which biological systems are responding across all continents. The United Nations Intergovernmental Panel on Climate Change, an alliance of over 2000 scientists from 100 nations, came to this conclusion. The panel concluded in 2001 that people were mostly responsible for these changes, citing deforestation and the use of fossil fuels, which released gases like carbon dioxide that trap heat. The rate of carbon dioxide buildup and atmospheric warming is accelerating; polar and alpine ice is melting at rates that were previously thought to be unlikely; the deep ocean is warming up and circumpolar currents are speeding up; and warming in the lower atmosphere is delaying the restoration of the stratospheric ozone shield.

Heat waves, such as the one that struck Chicago in 1995 and left thousands hospitalized and dead, have become increasingly frequent(Lagadec, 2004). The impacts are amplified by hot evenings, which are becoming common due to global warming. Long-term climate change can be harmful to human health. Due in part to variables associated with climate change, the prevalence of asthma in the United States has doubled over the last 20 years. Respiratory aggravations for residents of Caribbean islands originate from dust clouds that travel across the Atlantic Ocean by trade winds that are accelerated by increasing pressure gradients over warming waters. The dust clouds originate in African deserts. Elevated concentrations of soil fungus and plant pollen could also be implicated. Certain soil fungi are developed and sporulate by elevated carbon dioxide levels, and diesel molecules/droplets aid in the transportation of these aeroallergens deep into our alveoli and their presentation to immune cells along the way. The ice cover melting process is already causing physical trauma. Severe winter storms and risky travel conditions are another effect of ocean warming and Arctic melting in the continental United States. Shifting oceanic pressures cause winds and storms to intensify. The rise of infectious diseases appears to have been facilitated, at least in part, by changes in biodiversity and growing socioeconomic disparities over the past three decades. These changes have affected the balance between predators, competitors, and prey, which helps control pests and infections. The range of bacteria and vectors is limited by temperature, and the timing and intensity of disease outbreaks are influenced by weather patterns, which contribute to the transmission of these diseases. As the winters get warmer, disease-carrying ticks in Sweden are migrating northward, and models predict a similar shift in the US and Canada. Mosquitoes are extremely sensitive to temperature fluctuations and are known to carry a variety of diseases. Warming of their surroundings increases their rates of reproduction, extends their breeding season, and accelerates the maturation of the microorganisms they spread. As permafrost thaws and glaciers disappear in some locations, plant communities and insects are moving to higher ground. The West Nile virus (Figure 1.7) and hantavirus pulmonary syndrome were brought to the United States by the heightened weather variability that comes with climatic instability. Diseases that impact agriculture, forests, animals, wildlife, and marine life pose an even bigger risk to human health. The resultant biological deprivation could have significant effects on our food, water, and air.

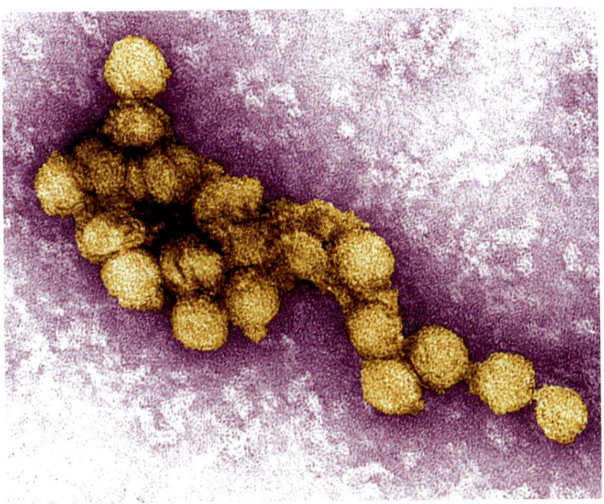

Figure 1.7 West Nile virus

Source: https://en.wikipedia.org/wiki/West_Nile_virus#/media/File:West_Nile_Virus_Image.jpg, *Creative Commons License.*

1.7 UNPREDICTABLE WEATHER

More unpredictable weather, disappearing pollinators, and an increase in pests and diseases are all challenges. Hurricane Ivan brought one fungal disease (soybean rot) into the country. Moisture and warmth will encourage its spread. Furthermore, a lot of habitats are struggling. For instance, our coastal zones are under threat from fungal and bacterial diseases, overfishing, excess waste, physical damage, and warming-induced bleaching of the coral reefs. Reefs preserve fish, which are a major source of protein for many island residents, and provide a barrier against storms and salinization of the groundwater. The cone snail generates a peptide that is 1000 times more potent than morphine and does not cause addiction (Bingham et al., 2010). It is possible that our understanding of the range of biological reactions to climatic shifts is incomplete. Preparation is necessary for the treatment of climate-related diseases, and early-warning systems that predict extreme weather can help limit the number of victims and slow the spread of disease. However, primary prevention would necessitate stopping the mining, extraction, transportation, and burning of fossil fuels, a change that many scientists feel would have numerous positive effects on the environment/human health in addition to aiding in climate stabilization. The financial advantages of the shift to renewable energy may also be underestimated. Renewable energy, energy-efficient, and hybrid technology, green buildings, and improved public transit networks can all be 21st-century growth engines when the associated financial incentives are sufficient. Warm sea surfaces evaporate quickly, and as the deep ocean warms, the water that replaces the evaporated material is likewise warm, feeding storms that come after. The series of megastorms is a sign of climate change. This could also be a turning point in our comprehension of how the climate is changing so quickly and how the climate system is magnifying natural phenomena. Large-scale opportunities are also made possible by prevention, recovery, and adaptation. Building a diverse portfolio of safe, evenly distributed, and environmentally friendly energy sources, along with technologies, can support economic growth, improve public health, strengthen energy security, and aid in climate stabilization.

1.8 ENVIRONMENTAL FACTORS

Environmental factors have varying effects on different infectious diseases. Naturally, while any direction of climate change may lead to a rise in the prevalence of one disease, it may also result in a decrease in the incidence of other diseases. Depending on the specifics of the local infectious disease situation, the overall consequences on human health for any given geographic location could be good, neutral, or negative. A deeper comprehension of how global warming affects certain infectious diseases will be necessary to make predictions about the overall effects of the phenomenon. Dengue and influenza are two deadly infectious diseases that have the potential to spread worldwide.

Dengue is a warm-weather disease, while influenza is a cool-weather one, both brought on by encapsulated RNA viruses. An analysis of the incidence curves in the US, Mexico, Colombia, Brazil, and Argentina reveals a strong dependence on the timing of annual influenza incidence. In high-latitude countries, epidemic peaks occur throughout the winter; in countries closer to the equator, there is no discernible annual peak. Many theories have been put forth to explain why influenza has a strong winter seasonality in addition to seasonal variations in ambient temperature. One set of reasons is changed host immunity, which may result from low vitamin D from less sun exposure or from changes in the photoperiod with fewer days in winter. Another set of explanations centers on heightened human-to-human interaction during the winter months, such as heightened social interactions among kids throughout the regular school year or breathing in confined areas while sharing air. It is plausible that multiple factors influence the seasonality of influenza. The impact of changed environmental variables on influenza transmission from animal to animal has been better understood due to experiments conducted on laboratory animals. Researchers demonstrated more than 40 years ago that low temperature and low relative humidity can enhance influenza transmission when they studied the spread of the disease in mice(Jaakkola et al., 2014). Dynamic systems, like epidemics, are frequently made up of two or more semi-independent but partially interacting dynamical subsets like immunity, environmental conditions, and health systems. To comprehend the impacts of a given factor, like ambient temperature, it is crucial to separate and examine these dynamical subsystems.

One typical strategy for controlling plant disease is a pesticide. It is now unfeasible in many systems to produce crops without pesticides for some plant diseases, like potato late blight. Some farmers and their families suffer from chronic pesticide exposure in areas where pesticide safety education is deficient. An estimated 1 to 5 million incidents of pesticide poisoning occur annually, with thousands of deaths associated with it(Karunarathne et al., 2020). Changes in the need for pesticides owing to the climate or other conditions may also have unforeseen effects on human health. This is because changes in the usage of pesticides may result in variations in unmanaged pesticide exposure. For some diseases, additional disease management techniques include eliminating contaminated plant materials, applying biocontrol agents, managing soils that suppress disease, or using certified seed to prevent pathogen introduction may be helpful. In agricultural systems, deploying genes resistant to disease is frequently the most interesting strategy for managing disease. While effective resistance is recognized for some diseases, partial resistance may nevertheless be a valuable component. For certain diseases, resistance provides effective management. Growers and consumers bear small expenses from the adoption of resistance genes, with the possible exception that plant breeders may find it difficult to integrate desired resistance genes with other desirable plant features in some situations. Finding persistent resistance presents another difficulty in the process of breeding crops for disease resistance. When resistance genes can effectively combat huge pathogen populations over an extended period, even when subjected to disease-promoting environmental conditions, then the deployment of those genes is considerably more efficient. The management of many diseases is reduced by the persistent issue of pathogen adaptation to overcome disease resistance. Given that the shorter resistant varieties surround the susceptible varieties, it appears that microclimate plays a significant role in this situation.

In some ways, managing plant diseases in agriculture is simpler than managing human diseases since farmers can modify the genetic makeup of food plants. This potential, however, can also cause issues if certain crop varieties that are very effective proliferate widely. Thus, the general homogeneity of cropping systems in the United States and trends toward increasing crop homogeneity in most regions of the world present a similar issue for plant disease management. Plant diseases that are suited to the common crop types can spread more quickly among agricultural plant populations due to this homogeneity. An indicator of how easily an organism may travel across a landscape is its connectedness. There is a close connection between maize and soybean. Because the production of cotton and wheat is more dispersed, disease populations are unable to spread quickly throughout all of the producing areas. Crop homogeneity could rise with conversion to biofuel production. With accurate and broadly useful forecasting models, disease risk maps based on climate can be produced. For instance, researchers used climate characteristics to map the risk of potato late blight. These models are limited to highly researched diseases; however, some researchers have created a general model of infection risk that can be used to map the risk of novel pathogens for which comprehensive models are not yet available. To assess how global risk is changing in response to climate change, models of pathogen risk can be combined with maps showing the current and projected climatic conditions.

Chapter 2

GLOBAL WARMING

Contents

2.1 INTRODUCTION

The twentieth century witnessed an unparalleled surge in both the human population and the need for carbon-based fuels. Despite restraints, the world in the twenty-first century appears incapable of managing the carbon emissions. Although there is no doubt that climate change will have an influence on human health, the effects of global warming (Figure 2.1) interact with one another to either amplify or lessen the risks to human health.

Figure 2.1 Global warming

Source: https://science.nasa.gov/wp-content/uploads/2023/11/Effects_page_triptych.jpeg?w=1280&format=webp, *Creative Commons License.*

Human health is at growing risk due to the multiple climatic effects of global warming. For instance, Hurricane Katrina's impact on the US Gulf Coast proved to be less of an issue than expected(Cutter et al., 2006). Rather, many factors, including environmental deterioration, inadequate preparation for disasters, and weak barriers to protection, combined to create one of the prime disasters in American history. In fact, researchers have indicated some possibilities, some of which are favorable, for the health of humanity in the future on a gradually warming planet. For instance, the likelihood of someone dying in the winter from exposure to low temperatures will be much lower. Furthermore, extended growing seasons in some regions will promote human health through a healthy diet. However, the World Health Organization (WHO), the Pew Center on Global Climate Change, and the Intergovernmental Panel on Climate Change (IPCC) concur that, after analyzing all the data, the detrimental effects of climate change will outweigh the positive effects on global health. In 2002, the combined causes of diarrhea, malaria, and protein malnutrition accounted for around 3.3 million deaths worldwide(Müller & Jahn, 2009). Human health will be impacted by global warming in both direct and indirect ways. The immediate temperature consequences of global warming will be the most noticeable. Heat waves occur more frequently as temperatures rise. The combined effect of these conditions points to an increase in the number of deaths from heat stroke. Furthermore, heat waves in cities are likely to be worsened by the urban heat island effect, raising the temperature in comparison to rural places. People are at greater risk of health problems from direct temperature effects in an increasingly urbanized society.

Additionally, scientists estimate that the frequency and intensity of droughts will probably rise due to global warming. Severe droughts can have disastrous effects on human health in a number of ways. Water shortages are first caused by a lack of rain. Food shortages are caused by reduced crop yields and livestock fatalities as a result of this condition. People are more prone to experience malnutrition and famine when they do not have access to a large range of foods or enough food overall. Weakened by starvation, their bodies are also unable to fight off infectious infections. Additionally, more and more uncontrollable wildfires are directly caused by drought. Apart from the immediate risk of burning, wildfires also have indirect health effects due to air pollution, destruction of homes and crops, and occasionally, forced evacuations. People start to feel the effects of heat stress at higher temperatures, especially when there is significant humidity, which hinders the body's capacity to cool itself. Extreme cold exposure without proper clothing can cause mild to severe hypothermia, which can be fatal, as well as a reduction in body temperature. Furthermore, those who are wet have a far lower tolerance for cold because moisture transfers heat away from the body. Human mortality and morbidity rise when temperatures hit severe levels, whether they are hot or cold. This is especially true the longer the extreme temperatures persist. Extreme temperatures cause more diseases and fatalities in humans. Furthermore, as temperatures rise and fall, the number of deaths also rises sharply with each degree.

2.2 HEAT EXTREMES

The incidence of extreme heat waves is expected to rise in the coming decades. Average temperatures are rising, but extreme temperatures are rising even more. Hot days and nights have grown more frequent, and cold days and nights have grown less frequent in recent decades. Heat waves and high-temperature extremes have become more common.

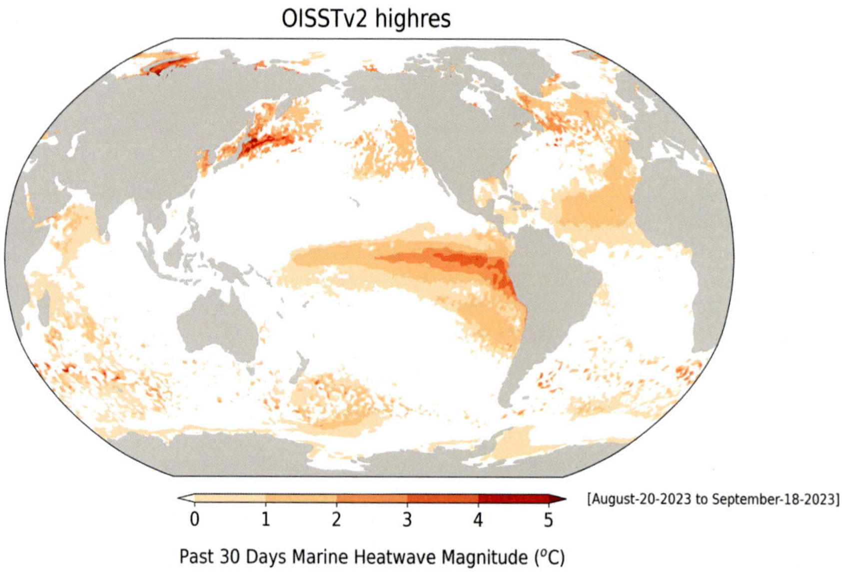

Figure 2.2 Heatwave

Source: https://en.wikipedia.org/wiki/Marine_heatwave#/media/File:Recent_marine_heatwaves.png, *Creative Commons License.*

A heatwave (Figure 2.2) is merely a period when the average daily temperature climbs far higher than what is typical for the season. It goes without saying that these conditions are especially unsafe in the summer when temperatures are already high on average throughout much of the planet. There are three types of heat waves: mild, moderate, and severe. Higher temperatures continue for longer periods during severe heat waves. The body's capacity to withstand heat decreases quickly with time. The population is becoming more and more concentrated in metropolitan regions. Cities frequently produce heat islands or environments of their own. The heat island effect can cause city temperatures to rise by 2° to 9°F, or roughly 5°C(Mohajerani et al., 2017). There is comparatively less vegetation in metropolitan areas, especially those with high human densities, than in more rural locations. Plants and trees help to mitigate excessive heat. Shading from trees and other plants keeps buildings cooler and maximizes the efficiency of air conditioners. Evapotranspiration, a process that trees and other vegetation use to cool the air, is reported by the U.S. Environmental Protection Agency (EPA). This process involves trees absorbing water via their roots and evaporating it through their leaves, which releases part of the heat trapped in the atmosphere. During a heat wave, cities get significantly hotter than rural regions because they have fewer trees and plants. The rise in impermeable land covers, such as concrete parking lots, buildings, and roadways, is a factor conducive to the urban heat island effect. These constructions stop precipitation from penetrating the soil and subsequently evaporating, which would have cooled the surrounding air. Instead of seeping into the ground, precipitation swiftly flows off the impermeable surfaces and enters drains. Many automobiles and industries in cities contribute significantly to the amount of carbon dioxide they emit. One greenhouse gas that keeps heat close to the surface instead of allowing it to escape into the upper atmosphere is carbon dioxide. For these reasons, a mild heat wave may become potentially fatal for urban people due to the urban heat island effect. The urban heat island effect is expected to have a detrimental impact on human health as the number of extreme heat events predicted as a result of global warming surges. The effects of more heat waves on health are compounded by an older population. In many parts of northern Europe, North America, and Japan, the median age of the population is rising. Public health experts have long recognized that young children and the older individuals are more susceptible to the negative effects of heat exposure. Increasing the number of older adults in the population will increase the quantity of the population at risk because a decreased ability to thermoregulate is a normal part of the aging process. Older people are far more susceptible to the harmful effects of heat because they are less competent than younger people to regulate their body temperatures.

2.3 DROUGHT

Up to 52,000 people died as a result of the heat wave that struck Europe (2003) (Kondratyev et al., 2006). Following that, the French government implemented several actions that have since been found to effectively save lives. These include the implementation of cold shelters where individuals can go to lower their body temperatures and health warning systems. This is an example of evaluating and adapting to personal risk. Similarly, in extreme heat events, the CDC in the United States has developed several techniques to help reduce disease and death. Despite the health risks posed by the direct temperature effects of global warming, which can result in fatalities and chronic diseases, public health officials are making great efforts to get ready for the increased frequency of heat waves that almost all models predict. El Nino occurrences, a warming cycle in the eastern Pacific Ocean, have an impact on monsoons, those seasonal changes in wind direction that typically bring rainfall. The monsoons are weak and do not bring the rain needed to survive during El Nino years. El Nino episodes are expected to occur more frequently in the coming years due to global warming. The health of millions of people will be negatively impacted by crop failures and drought as a result of this shift. A drought brought on by El Nino devastated Indonesia, resulting in massive food losses. A very wet winter ended the drought, but it brought further misery because planting could not begin in time, and rats and locusts multiplied, decimating stored food stores as well as crops. Due to this succession of poor harvests, many families were forced to eat less rice and turn to the less nutritious alternative of dried cassava.

Figure 2.3 Drought

Source: https://en.wikipedia.org/wiki/Drought#/media/File:Corn_shows_the_affect_of_drought.jpg, *Creative Commons License.*

Drought (Figure 2.3) does not cause famine, although it does play a role in the gradual development of famine conditions. Famine is both a slow-developing disaster and a process. People who live in dry regions start to experience health problems due to drought long before they start to starve to death. The process begins with low yields and crop failure. There are then severe food shortages and sharp price increases. Individuals with lower incomes tend to limit their food intake and reduce the variety of meals they eat by opting for less expensive, high-calorie items that may not be as nutrient-dense. Variability and change in the climate are linked to both acute and chronic nutritional problems. Drought has an adverse effect on health, including respiratory and infectious disorders, undernutrition, protein malnutrition, and/or deficits in some micronutrients. Under nutrition is the outcome of consuming less food. Protein malnutrition results from consuming insufficient amounts of protein, either from plants or animals. The body is experiencing micronutrient shortages due to a deficiency in critical nutrients, which can be found in a range of meals. The neurological condition beriberi is caused by low thiamine (vitamin B_1). All these disorders are associated with reduced nutrient diversity; when individuals are deprived of a diverse range of foods, they experience malnourishment on one or more fronts.

Drought is especially risky in Africa, where more than forty-three percent of the land is made up of arid regions(Waha et al., 2018). Famine brought on by drought has killed more people in Africa in recent years than in any other continent. Drought is the most significant natural hazard in Africa in terms of destroyed livelihoods, famine, fatalities, and diseases linked to malnutrition. When a drought is brief, both people and areas recover rather quickly. However, when droughts last for years, they get worse very quickly. Chronically malnourished individuals are far more vulnerable to both nutrition-related and infectious disorders. Furthermore, AIDS may exacerbate the negative nutritional effects of drought. This creates a lethal loop whereby those who are undernourished are unable to resist diseases, and those who are already afflicted with a serious disease are more susceptible to being malnourished during dry spells. While starvation and malnourishment are often the most hazardous health effects of drought in developing nations, there is another serious health risk associated with global warming and drought in regions such as North America, Europe, and Australia. Based on data from the last few decades, climate change will cause increased fire activity. There are several detrimental impacts of wildfires on health. Swiftly spreading flames, driven by wind through dry vegetation, have the potential to engulf humans and animals, causing severe burn injuries or even death. Hurricanes, typhoons, or cyclones in the region will intensify into strong tropical cyclones with increased winds, torrential rainfall, and destructive storm surges.

2.4 RAINFALL

There is a chance for extreme rainfall events and larger storms, when locations get a lot more rain in a shorter amount of time than in the past. Furthermore, researchers continue to examine the causes underlying El Nino, an intermittent warming of the Pacific Ocean that modifies global weather patterns. While there is disagreement on the relationship between global warming and El Nino or La Nina, which is a cooling of Pacific water temperatures that frequently occurs after an El Nino year, most climatologists concur that global warming will exacerbate El Nino's effects on regional weather. A few scientists support this assertion by using El Nino during the 1997–1998 period, pointing out that 1998 was the hottest year on record due to the shared effects of global warming and El Nino. That year, it caused $100 billion damage worldwide, leaving catastrophic floods along the Yangtze in China, massive fires in Indonesian Borneo and the Brazilian Amazon, and extreme drought in Mexico and Central America. It was also the strongest El Nino in 130,000 years. Some contend that when El Nino's impacts compound with those of global warming, weather extremes can become disastrous. The World Health Organization (WHO) notes a significant risk to human health from weather extremes, particularly from storms, floods, and heavy rains. According to the WHO, weather-related disasters claimed the lives of about 600,000 people globally in the 1990s(Paul, 2020). Among the health concerns associated with global warming that are most hazardous are hurricanes and floods brought on by excessive rainfall. Anyone in the path of a hurricane wind could be killed or seriously injured. During a hurricane, when a window breaks, the wind rushes in, raising the roof, and the house collapses on top of itself, smashing possessions and hurting or killing anyone who hasnot escaped. Furthermore, large objects like tree branches and construction materials can be lifted and carried by powerful winds. These flying objects are similar to deadly weapons; during a storm, unprotected individuals could be struck by branches, boards, nails, or other debris that the wind has carried into the air.

Figure 2.4 Flooding

Source: https://en.wikipedia.org/wiki/Flood#/media/File:Katrina-new-orleans-flooding3-2005.jpg, *Creative Commons License.*

Water is far more harmful than wind. Storm surges and flooding (Figure 2.4) cause more deaths than any other type of natural disaster. Floods occur when rivers exceed their banks due to heavy rains or when the land becomes so saturated that it can no longer hold water. Heavy rains can occasionally result in flash flooding, by overflowing a dam or by forcing water down a small valley. There is a slight warning to those who are caught in the path of a flood; the fortunate ones find themselves clinging to trees while they wait for help. Flash floods will occur more frequently as a result of the increase in intense rain events brought on by global warming. Extreme rain events that struck a large area of the midwestern United States in 1994 and 2008 caused catastrophic floods (Retchless et al., 2014). In addition, people across the world have been impacted by significant flooding caused by intense rainstorms between 1990 and 2009. For instance, in 2003, floods in China caused 130 million people to be displaced, or dead. In addition, Manila, the capital of the Philippines, received 16.7 inches (43 cm) of rain in just twelve hours on September 27, 2009, due to Tropical Storm Ketsana. It also produced severe flooding, which resulted in the deaths of over 100 people and the displacement of almost 280,000 people. Furthermore, a large number of people drown every year when they try to move across a flooded area. Few people are aware that a car can be completely destroyed in minutes by flowing water only 6 inches (15 cm) deep, submerging drivers and passengers in floodwaters. Storm surges are also produced by hurricanes. When tropical cyclonic winds force water ashore, it creates a storm surge. This surge and the approaching tide cause the water level to rise sharply, by as much as 15 to 25 feet (4.5 to 7.6 meters). In addition to the surge, strong storms with strong winds also produce giant waves.

2.5 WEATHER EVENTS

Water can therefore quickly build up on the coast, ruining houses, washing away cars, and injuring or killing those who drown. The storm surge has the greatest potential for loss of life related to a hurricane, according to the NOAA. Mudslides are a typical aftereffect of tropical storms (heavy rains) in several parts of the world. When steep hillsides are cleared of trees and other vegetation, severe rains can cause the mud to slide down the hills, destroying people and houses in its path. Thirty thousand people died in Venezuela in 1999 as a result of landslides and floods that followed storms(Blanchard-Boehm, 2004). There are more indirect hazards associated with extreme weather, even though the direct health risks resulting from global warming may seem obvious. When a powerful cyclone like Katrina or Nargis strikes an area, it can lead to a series of dangerous situations. In fact, these storms can occasionally result in an infrastructure failure. Emergency and medical vehicles cannot reach the injured or dead when roads are obstructed by trees or washed out. Rescue turns challenging or unfeasible. Some people who have survived the first wave of severe weather eventually lose their lives because rescue efforts are delayed. During floods and storms, there are also indirect health risks besides infrastructure collapse. Floodwaters can get contaminated with hazardous substances like chemicals, heavy metals, or oils as they pass through industrial regions. Both rescue workers who come into contact with the toxic water and people who have to walk or swim through these waters to get to safety run the risk of becoming sick. The potential for mass human exposure to hazardous compounds discharged during disasters increases with increases in population density and accelerated industrial expansion in areas prone to natural disasters. Power lines that come down and gas lines that break might be caused by extreme weather. Again, in industrial regions, ruptured gas lines can result in wildfires that spill harmful chemicals into the air, injuring anyone who inhales the fumes. Furthermore, damaged electrical wires are a serious risk to human safety, especially if they are submerged in water. Both rescue workers and victims of hurricanes and floods run the risk of electrocution.

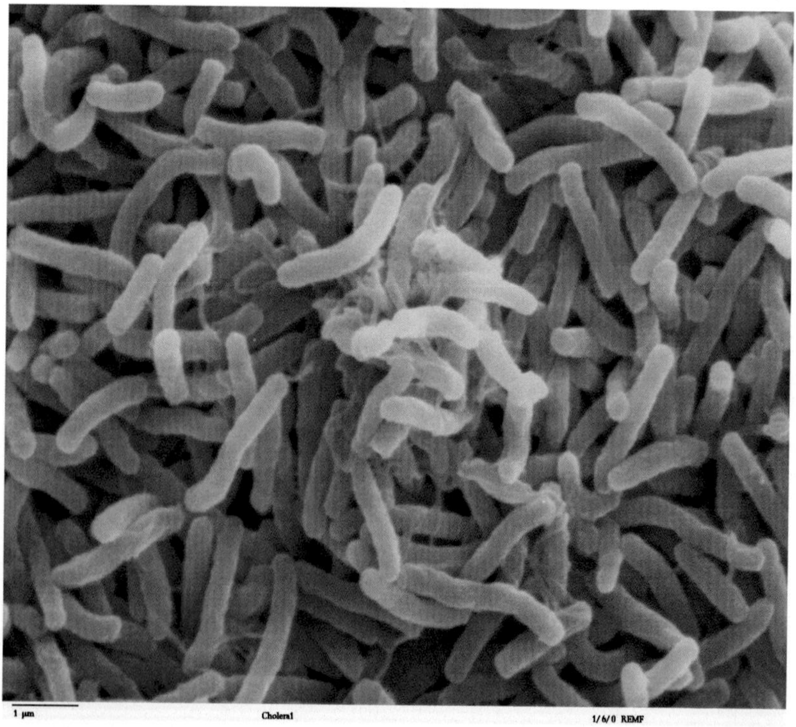

Figure 2.5 Cholera

Source: https://simple.wikipedia.org/wiki/Cholera#/media/File:Cholera_bacteria_SEM.jpg, *Creative Commons License.*

Following severe weather, freshwater contamination poses a major risk to public health. People must boil their water before drinking since sewage runoffs carrying feces frequently contaminate drinking water sources. When there are power outages, people may not have access to sterilize their water, forcing them to choose between dying from thirst and contracting a water-borne disease. Food reserves also get contaminated over time, either by flooding or by not having refrigeration because of power outages. The bacterium *Vibrio cholerae*(Figure 2.5) frequently causes widespread outbreaks of the cholerawhen water and food supplies are contaminated. Cholera is characterized by severe diarrhea and dehydration. The Mayo Clinic states that cholera can soon become lethal if treatment is not received. Months or even years may pass after an extreme weather event before the indirect health effects become apparent. Standing/stagnant water left after flooding gives mosquitoes, some of which are capable of carrying malaria, the ideal breeding environment. Malaria mosquitoes are predicted to migrate further north in the upcoming decades because of climate change, endangering the health of those who experience floods in higher latitudes. Similarly, water and moisture after floodwaters recede, especially in a warming climate, increase the risk of mold infection in homes. The two ways to prepare for extreme weather events include improving the storm-worthiness of homes in tropical storm-prone locations and creating disaster alert systems to notify locals when a dangerous storm is approaching. Local and regional governments need to create effective evacuation plans to guarantee a safe departure from regions under risk. Furthermore, better sanitation and sewage disposal systems reduce the possibility that human waste would pollute floodwater, loweringthe risk of cholera and other water-borne diseases developing. Preserving human health can also be achieved by identifying floodplains and blocking human development. It is also in the best interests of human health to make sure that infrastructure can resist strong winds and heavy rainstorms.

By quickly reaching hurricane and flood victims, skilled rescue personnel and emergency responders can save a great number of lives. The function that ozone plays in human health and disease must be examined before delving into any topic of global warming and air quality. The element of air that is most crucial to humans is oxygen (O_2), which is a gas made up of two oxygen atoms. Ozone (O_3) is a gas made up of three O atoms. The National Aeronautics and Space Administration or NASA reports that life on the planet is protectedbecause stratospheric ozone shields Earth's surface from the sun's ultraviolet (UV) rays. The U.S. Environmental Protection Agency (EPA) claims that ozone accomplishes this by lowering human exposure to UV radiation that causes skin cancer and cataracts. About 90% of the ozone in the atmosphere is found in the stratosphere, which is located 5–30 miles (8–48 km) above the surface of the earth(Junge, 1962). There, ozone is created by sunlight and destroyed by certain chemical reactions, some of which are caused by humans. Because stratospheric ozone is vital to life as we know it, scientists, policymakers, and the general public are quite concerned when they see gaps in the ozone layer. The troposphere, the part of the atmosphere nearest to the surface of the earth, also contains ozone. The Royal Society states that sunlight combines with naturally occurring chemicals and contaminants in the air to generate ozone. These substances are produced by forest fires and car exhaust emissions. When the air is stagnant and the weather is hot, there is a high production of ozone. These days, the EPA directs citizens to reduce ozone pollution by driving less. People should safeguard their health, as well as the health of their children and other family members who may be suffering from long-term respiratory conditions like asthma. Because of global warming, these situations are becoming more likely in several parts of the world.

2.6 OZONE

High ozone concentrations can be harmful to health if inhaled. According to NASA, reactions to ozone include shortness of breath, wheezing, and occasionally even nausea. Ozone causes inflammation by damaging the lining of the human throat and lungs when it gets inside. According to the EPA, there are significant health concerns associated with spending eight hours or more exposed to ozone levels above 80 parts per billion, including irritation of the throat and lungs, asthma attacks, and emphysema. Even in healthy individuals, ozone may harm lung tissue. In fact, when ozone levels are at 80 parts per billion, those who exercise, even for brief periods, have a diminished capacity to breathe. Smog is formed by fusing smoke and fog. Steel plant chimneys frequently emitted a dense smoke that enveloped areas like Pittsburgh, Pennsylvania, and Newcastle, England. Particularly in the summer, these emissions heat the atmosphere and create heavy smog clouds that are heavy in ozone and other pollutants. The air quality is deteriorating due to haze and particulate pollution as developing nations burn fossil fuels and utilize chemicals, becoming more industrialized and urbanized than developed countries. Poor air quality has been recognized as an undesirable side effect of urban population concentrations and intensive industrial and agricultural activities.

2.7 SMOG

Smog is more common due to increased climatic conditions brought on by global warming(Mickley, 2007). For instance, smog is trapped near to the earth during atmospheric inversions, which are meteorological phenomena in which the air temperature rises with altitude rather than falls. In an inversion, the air closest to the ground is heavier and colder than the extremely hot air above, preventing it from rising and causing breezes that would otherwise disperse the smog. Thus, many urban locations, like Mumbai, India, and Los Angeles, California, experience air stagnation due to hot, dry summers with frequent atmospheric inversions, carrying all the health risks and hazards associated with elevated ozone levels. Particulate matter can also be found in smog, while some types of particulate matter can circulate without smog in both rural and urban locations. Any microscopic material that floats in the air is referred to as particulate matter. These particles are mostly undetectable. They may not be visible, but they nonetheless pose the greatest health hazards associated with lower air quality. There are several sources of particulate matter. Particles might originate from factories, farms, smoke, and dust from dirt roads. They are frequently produced by crushing rocks or disturbing the soil, which releases particles that are then carried by the wind. Fires, the smelting and processing of metal, and car exhaust all release some extremely small particles into the air. These particles may consist of heavy metals like mercury or hazardous substances. Certain particulate matter is actually liquid in the form of microscopic droplets that are occasionally created by industrial operations and can include compounds that are harmful to the lungs. Furthermore, emissions from power plants are a significant source of all forms of airborne particulate matter. The need for power plants to generate electricity for air conditioning and refrigeration will increase as the earth heats, and this need may lead to an increase in the quantity of particulate matter that these plants emit. There are many ways in which chemicals and diseases from agriculture might come into contact with humans without their knowledge. Humans may be exposed to agriculturally generated chemicals and pathogens in the environment, i.e., air, soil, water, and sediment through a variety of channels. One way to become exposed is via breathing in particles. Overall, climate change is likely to increase human exposure to agricultural contaminants.

2.8 WILDFIRE

Wildfire frequency is predicted to rise in tandem with global warming(Baker, 2022). These fires release ash and soot into the atmosphere, which concentrates particulate matter and lowers air quality. In addition, hazardous compounds found in factories or building materials are included when wildfires spread into residential neighborhoods or industrial areas. According to the EPA's projection, there could be a potential rise in the health issues linked to particulate pollution and smog. Exposure to air pollution can aggravate chronic respiratory and cardiovascular disease, damage lung tissue, lead to premature death, and may even contribute to cancer. Global warming might make these issues worse. Airborne particles impair the lung's defense against ozone depletion. Thus, lung tissue is harmed twice by the combination of ozone and particle pollution. Epidemiologists have consistently found significant increases in emergency-room (ER) admissions and cardiorespiratory deaths during episodes of high levels of both atmospheric ozone and particulates in several American and

European cities. The combined effect of the two pollutants was greater than the effects of each pollutant acting alone. Individuals who are subjected to elevated concentrations of particulate matter may experience irregular heartbeats, asthma attacks, and cardiovascular ailments. People who already have emphysema or chronic bronchitis may find that smog and particulate pollution make their diseases worse. Additionally, the body becomes more susceptible to attacks from infectious respiratory disorders like pneumonia and influenza due to the harm done to the lung lining by the combined impacts of ozone and particulate matter. The American Lung Association estimates that the annual number of asthma attacks caused by power plant emissions alone is approximately 366,000, and the annual number of deaths resulting from air pollution is between 200,000 and 570,000(Ghude et al., 2016). Any airborne material that triggers an allergic reaction in humans is referred to as an aero-allergen. Mold, spores, pollen, and fungi are a few examples. For a number of reasons, global warming may raise the concentration, geographic distribution, and incidence of aeroallergens (Figure 2.6).

Figure 2.6 Aeroallergens

Source: https://en.wikipedia.org/wiki/Allergen#/media/File:Misc_pollen.jpg, *Creative Commons License.*

2.9 ALLERGENS

Airborne allergens, however, can still occur in places that do not have dry weather favorable for plant pollen dispersal(Oh, 2018). Conditions that are warm, humid, and moist will encourage the growth of fungi and mold both indoors and outdoors. While the majority of molds are benign, some of them can create compounds called mycotoxins that are harmful to human health. Moreover, both adults and children may experience severe allergic reactions as a result of breathing in mold spores. The most common health problems due to mold exposure are running nose, sinus congestion, eye irritation, cough and congestion, sore throat, sneezing, upper respiratory infections, headaches, worsening asthma, and fatigue. Dangerous amounts of mold-infested the homes since the residences are typically well-insulated and there was minimal air circulation. Many inhabitants became ill because of the mold spores that were later discharged into the air within the homes or businesses. Hazardous materials professionals had to clean a lot of homes; to get rid of the mold, some had to be entirely destroyed. The likelihood of mold infestation remains a health concern since predictions from global warming models indicate that some regions would experience more precipitation throughout the winter months. Due to global warming, these conditions will occur more frequently and will promote the spread of potentially harmful fungi into new areas.

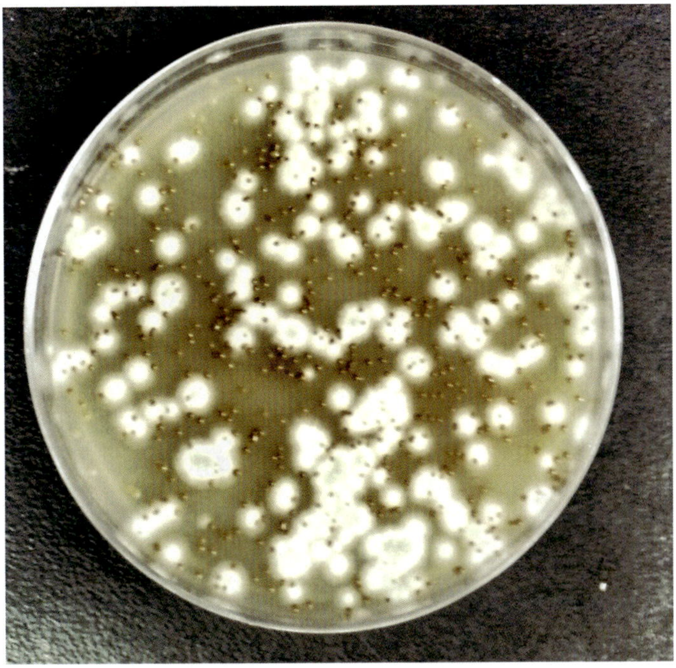

Figure 2.7 *Cryptococcus gattii*

Source: https://upload.wikimedia.org/wikipedia/commons/2/2f/Cryptococcus_gattii_selection.jpg, *Creative Commons License.*

One such is the microscopic tropical fungus *Cryptococcus gattii*, (Figure 2.7), recently been found in the Pacific Northwest and British Columbia(Dixit et al., 2009). The inhalation of airborne fungal spores can cause people to contract cryptococcosis, a potentially fatal disease of the lungs and, less frequently, the central nervous system. Both healthy persons and those with weakened immune systems, such as HIV patients are susceptible to the fungus-related disorders. Prolonged coughing, excruciating headaches, fever, night sweats, and weight loss are common symptoms. Among the infected, skin lesions are also frequent. Although antifungal medications can effectively treat patients, the disease can be hard to identify because symptoms frequently take two months or longer to manifest.

2.10 AIR QUALITY

The human and animal populations will continue to face several air quality problems due to the inability to stop climate change effectively. For instance, ozone generation is probably going to continue as long as emissions are high and temperatures rise. Similarly, due to longer growing seasons and greater dispersal, airborne allergens will still be a concern to allergy patients. However, there are many ways to enhance the quality of both indoor and outdoor air, actions that will protect the lungs. First, make parents and schools keep kids inside on days when ozone levels are higher than 80 parts per billion. Media like radio and television can help with this effort, similar to how they publicize other weather-related occurrences, like winter weather advisories and severe thunderstorm warnings. Despite global warming, governments can take steps to guarantee that people can breathe clean air. Air quality will be enhanced by regulations on emissions from companies, power plants, and cars since they will lower particulate matter and ozone levels. Many policymakers and specialists on climate change contend that rich governments need to figure out how to assist developing countries in setting up ecofriendly factories and power plants. However, there is debate about this idea of governmental regulation all across the world. People in industrialized countries have grown accustomed to the conveniences that come with owning a car, air conditioning, central heating, and other possessions. It might be challenging to give up financial luxuries for the possibility of cleaner air, which the average person cannot observe or quantify. Furthermore, enterprises in industrialized countries are already bound by a plethora of rules concerning waste products and emissions. Many business owners complain about the arduous laws that are in place and argue that adding more will make their companies unprofitable. Developing countries aspire to become full partners in the global economy and build robust manufacturing bases. These countries frequently construct factories and power plants without air safety equipment because doing so would significantly increase the cost of industrial operations. In addition, people in developing nations desire vehicles and other items to make their lives more comfortable, much like their counterparts in wealthy nations. The stress that the processes that produce these goods place on air quality rises sharply as an increasing portion of the global population has access to them. Developing countries object to laws and regulations being imposed upon them by developed countries.

While medical experts are working hard to develop therapies for diseases brought on by airborne allergens like molds, fungi, and pollens, other scientists are investigating ways to lower air pollution levels directly. People should think about how their actions might contribute to the reduction of pollution, in addition to paying special attention to their own and their family's health. People should be aware of the signs of diseases brought on by or exacerbated by poor air quality, and they should collaborate with their physicians to get the best care available, particularly in cases of childhood asthma, which can be fatal. An agent known as a vector spreads an infection from one human or animal host to another. A zoonotic disease can spread from vertebrate animals to humans through a vector(Rahman et al., 2020). For instance, rabies can spread directly from an infected animal, such as a bat or fox, to humans.

2.11 TEMPERATURE ALTERATIONS

As the temperature increases and the ice melts, traditional food preparation and consumption also become riskier for indigenous peoples. Furthermore, the environment that warm sea waters provide is conducive to the development and spread of the bacteria that cause botulism, a dangerous and frequently fatal type of food poisoning. The native Arctic peoples are particularly vulnerable to this deadly disease because of their diet. Many Arctic inhabitants may eventually have to relocate as a result of melting ice. Many studies show that when compelled to migrate, indigenous peoples end up in urban areas where their health deteriorates. Researchers have found that compared to those tribes that have been allowed to stay on their territory, indigenous peoples who have been forced off their lands experience greater rates of death, shorter life expectancies, and more health issues. An alarming percentage of indigenous people commit suicide after being forced to leave their traditional settings due to alcoholism and mental health problems. In fact, a large portion of Inuit and other indigenous arctic civilizations are at risk of going extinct due to global warming. Everyone's health will be impacted by climate change, but not equally. Children are more susceptible to the health effects of global warming for various reasons. They must first rely on the adults in their lives to make emergency plans and preparations. If parents are not successful in this crucial duty, children will suffer. Extreme weather incidences frequently result in injuries to children, and occasionally the aftermath is a traumatic stress disorder. Moreover, because their lungs are still developing, children and newborns are particularly susceptible to the harmful effects of ozone and other pollutants. The environmental choices made by their parents, grandparents, and great-grandparents will likewise have an impact on the children. Another extremely sensitive group that will probably be affected more than most by the detrimental effects of global warming on health is the old people. Heat-related disease claims the lives of more elderly individuals than any other subgroup. The older individualsoften live alone and lack the resources necessary to prepare for or endure severe weather, heat waves, or droughts. They frequently already suffer from cardiovascular or pulmonary conditions, which are made worse by air pollution. Since their immune systems no longer shield them as completely as they did when they were younger, aged people are also the most likely to die from infectious diseases. Because of their physical fragility, they might not be able to access clean water and food, which could be scarce as aconsequence of global warming.

Poor people are more likely to experience heat stroke because they rarely have access to cooling systems during heat waves. They frequently reside in cities where the intensity of the heat is amplified. Many of the poorest people reside in coastal areas that are vulnerable to floods due to increasing sea levels and powerful storms(Felsenstein & Lichter, 2014). Many others may find it challenging to obtain food or water because they reside in rural areas that are affected by drought. When weather-related shortages occur, they frequently cannot afford the high prices, and their physical state is already compromised by disease and malnourishment. Additionally, they have the highest likelihood of becoming environmental refugees, that is, of being uprooted from their homes by droughts, floods, sea level rise, or melting glaciers. Poor nations lack the public health infrastructure necessary to sufficiently shield their citizens from the health risks posed by climate change. Furthermore, a large number of the poorest countries are situated in areas that will be hit by the effects of global warming.

Chapter 3

AGROECOSYSTEM ADAPTATIONS TO CLIMATE CHANGE

Contents

3.1 INTRODUCTION

The emission of greenhouse gases (Figure 3.1) is accelerated by both natural and man-made processes, including agricultural practices(Liu et al., 2019). Even though since the green revolution, some intensive agricultural practices such as the massive application of chemical fertilizers, burning of biomass, tillage practices, improper conversion of land use, deforestation, and other human activities have resulted in several greenhouse gases (GHGs), which cause the ecosystem to heat.

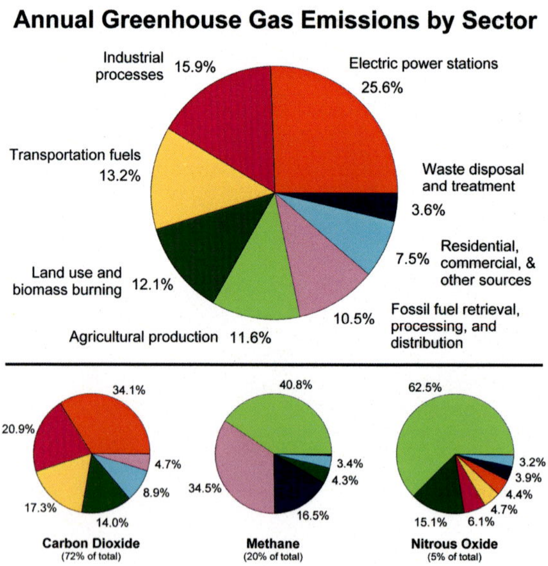

Figure 3.1 Greenhouse gases

Source: https://upload.wikimedia.org/wikipedia/commons/e/e0/Greenhouse_Gas_by_Sector.png, *Creative Commons License.*

This affects all living things, including biodiversity loss, the emergence of infectious diseases and insect pests, poor-quality timber, and the productivity of biomass and soil. The structure and functioning of agroecosystems are impacted by climate change (CC), which also impacts overall productivity and ecosystem services, all of which are necessary to support human and animal life. However, the agroecosystem itself contributes to greenhouse gas emissions by using chemical fertilizers, mechanized equipment, a significant number of pesticides and herbicides, and other practices while simultaneously releasing greenhouse gases into the environment. Climate-smart agriculture, location-specific agroforestry, conservation agriculture, non-tillage systems, integrated farming systems (IFS), etc., are therefore, practices that not only improve biomass productivity overall but also lessen the impact of climate change by increasing the capacity of soils and vegetation to store and sequester carbon. A healthy agroecosystem supports healthy soils and vegetation, in turn supporting all animals and people while preserving environmental security and food and nutritional security (FNS)(Hebinck & Oostindie, 2018). Consequently, there is a connection between environmental security, food and nutrition security, and agroecosystems. Effective government policy, non-governmental organizations, profitable organizations, and academic institutions should also identify some priority areas for research and development to improve the structure and functions of agroecosystems, which can lead to an improvement in biodiversity and ecosystem health.

3.2 INTENSIVE AGRICULTURE

Intensive agricultural practices today release several greenhouse gases (GHGs) into the atmosphere, with developing countries accounting for about three-quarters of global GHG direct emissions. Methane (CH_4) is released into the atmosphere by flooded rice cultivation methods. Nitrous oxide (N_2O) is released by the microbial conversion of nitrogenous fertilizers and manures; and carbon dioxide (CO_2) is released when biogeochemical cycles are altered, biomass is burned, and organic residues decompose. Similar to this, deforestation has been contributing to an increase in greenhouse gas levels and igniting the urgent concern that global warming is altering our climate.

Global agricultural practices, applying manure to soil and pasture/grasslands, applying synthetic fertilizer, cultivating rice, managing manure, and handling crop residues, among others, contribute to an estimated 4354 $MtCO_2eq$ of greenhouse gas emissions per year into the atmosphere. The Asian continent accounts for the majority of these emissions (45%). Anthropogenic activities such as deforestation contribute roughly 3500 $MtCO_2eq\ yr^{-1}$ GHGs to our atmosphere(Raj et al., 2020). Of these, America contributes the most (2100 $MtCO_2eq\ yr^{-1}$). Additionally, during the decade from 2000 to 2010, European countries contributed zero GHGs to the atmosphere. The agriculture sector contributes significantly to the growing population, but one of the main problems today is the 13.5% of greenhouse gases (GHGs) released from the agricultural sector as a result of improper land-use conversion, increased use of inorganic fertilizers, and rice cultivation. In addition to impacting overall food production, this can also accelerate desertification and land degradation; degrade soil structure and function as a result of poor soil quality; impact biodiversity generally; and create uncertainty on the security of food, health, and the environment.

3.3 AGROECOSYSTEMS

In this situation, natural ecosystems must be adapted to create more hospitable and sustainable agricultural systems that can meet human demands for both direct and indirect benefits as well as environmental health by reducing ongoing greenhouse gas emissions. In light of the vegetation and soil's ability for carbon sequestration (CS), climate-resilient agroecosystem techniques are therefore excellent methods for simultaneously improving soil and plant production. Agroforestry systems (AFs), conservation farming, climate-smart agriculture, no-tillage farming, and other techniques improve crop and vegetation diversity. They also improve soil structure and functions, which improves the landscape matrix. The structure and functions of the agroecosystem are affected by biotic and abiotic factors, including location-specific agricultural technology, its adoption and management practices, social and cultural contexts, and the environmental and biological conditions in the particular region. An agroecosystem is a modified and manipulated form of a natural ecosystem that is created through ecological intensification to improve farming systems that are more economical and productive for both humans and the surrounding environment. It also supplies a variety of vital ecosystem services. In contrast to conventional and intensive agricultural practices, conservation agriculture, climate-smart agriculture, agroforestry, integrated farming systems, organic farming, and the integration of microbial technology can help to enhance carbon absorption and retention in both soil and vegetation. This will lead to higher biomass and a reduction in continuous GHG (carbon emission) emissions, representing a more sustainable agroecosystem.

The scientific management of input elements, namely land, capital, and labor, is generally linked. Additionally, an analysis of the agroecosystem must consider three fundamental dimensions: the environment (or ecosystem) level, the economy, and the human element. These dimensions are necessary to thoroughly examine the intricacy of the agroecosystem on both a temporal and spatial scale. As a result, the study of an agroecosystem heavily depends on each component and how they interact with one another across various dimensions. The long-term phenomenon known as climate is rapidly changing because of many greenhouse gases (GHGs) being continuously released into the atmosphere by both natural and man-made sources. The most powerful gas responsible for climate change and global warming as a result of extreme weather events is CO_2, released as a result of industrial development, mining, deforestation, and the heavy application of chemical fertilizers in agricultural fields. For instance, the production of CH_4 from the growth of rice and the management of manure emitted 48% and 97% of GHGs in developing countries, respectively, compared to 52% in rich/developed countries(Vergé et al., 2007).

Similarly, burning biomass accounts for 92% of global greenhouse gas emissions from poor countries alone. Furthermore, the production of rice in South and East Asia accounts for 82% of global CH_4 emissions, while the burning of biomass in African and Latin American countries accounts for roughly 74% of emissions(Maraseni et al., 2018). Heavy application of chemical fertilizers instead of biological fertilizers and the flooding of fields under rice cultivation practices are examples of unsustainable agricultural practices that contribute to the emission of GHGs. These practices create anaerobic conditions that can enhance the CH_4 emissions by up to 90% in the atmosphere, and these emissions are regulated by both the morphological and physiological characteristics of rice crops. However, the use of machines, chemical, and inorganic fertilizers, pesticides, herbicides, and other agricultural practices, combined with the adoption of crops that exhibit high yields within an agroecosystem, results in the atmospheric release of greenhouse gases.

3.4 ALTERED PRODUCTIVITY

The use of large amounts of chemical fertilizers in agriculture during the green revolution era has undoubtedly increased productivity, but at the expense of degrading the environment by releasing greenhouse gases into the atmosphere. CO_2 is the most potent GHG that may be released from soil through biogeochemical transformations. Similar to this, the synthesis and conversion of nitrogen-based fertilizers by soil microbes results in the highest release of N_2O. Similar to this, arid and semi-arid agroecosystems have been subjected to intensive management practices as a result of the severe/rough environment and water shortage. These practices include higher soil disturbance through tillage practices, which emit 21% more CO_2 than tilled soil, and higher fertilizer application doses, which alter the properties of the soil and the environment and influence microbial activity, which may result in higher greenhouse gas emissions. Similar findings have been made about the declining emission values of CH_4, N_2O, and CO_2 resulting from intensive agroecosystem practices in the various land-use systems of the Indus Basin. Rice farming practices and intestinal fermentation are the main sources of CH_4 emissions; Organic matter (OM) mineralization and the microbial denitrification process are the sources of N_2O emissions; soil respiration, crop residue burning, and OM decomposition are the sources of CO_2 emissions. Because of this, the agroecosystem can function as either carbon emission or absorption, depending on the management and application types. The effects of climate change are well-recognized by developers, lawmakers, manufacturers, and farmers alike. The world is concerned about greenhouse gas emissions from both natural and human activities. These emissions impact not just our environment through global warming but also ecosystems on land and in the ocean. Severe weather may have an impact on wild animals as well as the morphology, reproductive biology, and species richness, diversity, and abundance of the flora. Agroecosystems refer to the intentional alteration of natural ecosystems through ecological intensification toimprove the services provided by a variety of outputs for the welfare of humans and cattle. The general structure and operations of the agroecosystem are impacted by the changing climate, which includes unpredictable rainfall, rising temperatures, and increased humidity. Drought and flooding are caused by variations in surface and ground water levels. The main effects of the changing climate and global warming are on FNS, human health, and land degradation and desertification.

Figure 3.2 Ecosystems

Source: https://en.wikipedia.org/wiki/Ecosystem#/media/File:River_gambia_Niokolokoba_National_Park.gif, *Creative Commons License.*

In addition to diversifying ecosystems, (Figure 3.2) the various approaches can potentially increase the amount of carbon in the soil and vegetation system. Thus, the steps towards adaptation and mitigation of extreme climatic events along with sustainable environments and ecosystems are climate-smart farming practices of agroecosystem along with desired ecological intensification and modification in the natural ecosystem for satisfying human needs in terms of tangible and intangible benefits. Improved agroecosystem management and practices lead to improved landscape structure and management through a variety of intermediate benefits, such as an increase in biomass and productivity that meets human food and nutrition needs, encourage the agroecosystem to produce both tangible and intangible benefits for the environment and people, and provide effective ecosystem services. Maintaining soil structure and essential nutrient availability is necessary for the health and productivity of both plants and cattle(Rayne & Aula, 2020). It is also important to improve biodiversity, which is crucial for a better landscape. Improving carbon storage and sequestration capacity in soils and vegetation helps mitigate global climate change problems. Improving pollination and breeding programs helps maintain progeny and overall population while reducing problems with land degradation and desertification. By optimizing the potential for storing and sequestering excess carbon from the atmosphere into the vegetation and soils, climate-smart agriculture, which is environmentally friendly and supportive of socioeconomic development, is a very effective strategy for mitigating extreme weather events. This can strengthen farmer income and livelihood and increase biomass productivity. Using climate-smart agricultural methods preserves soil, human health, and environmental security in addition to assisting in decreasing GHG emissions from the agricultural sector. Therefore, climate-smart agriculture achieves three goals: lowering greenhouse gas emissions from a variety of agricultural activities, increasing productivity and income that support national food security, and building resilience against climate change and global warming. Under its unique significance in soil and water conservation, plant and human health management through nutrient availability to plants and FNS, effective use and management of natural resources, improvement of livestock productivity and health, improvement of overall crop productivity and quality, reduction of excess greenhouse gas emissions into the atmosphere, and maintenance of ecological security and sustainability, conservation agriculture is a great strategy to combat the climate change problem.

3.5 AGRICULTURAL PRACTICES

Tillage minimization, yield diversification, maintaining vegetation throughout the year, adding plant residues to the soil to aid in the formation of organic carbon, and applying various organic supplements to the soil to maintain a population of healthy soil-inhabiting microorganisms are the main features of conservation agriculture. Accordingly, the best option for mitigating climate change is to use conservation agriculture good practices to increase the amount of atmospheric carbon that can be stored and sequestered in soils and vegetation. The no-tillage practice has a major impact on the soil because it increases the amount of organic carbon available in the soil and reduces CO_2 emissions into the atmosphere, which causes the climate change phenomenon. Due to soil organic carbon (SOC) loss from decomposition and erosion, soil disturbance from a variety of human activities and plowing techniques disrupts the soil and encourages the emission of CO_2. Many studies have been conducted on the benefits of tillage practices and their notable influence on augmenting the soil capacity to retain carbon. Similar to this, non-tillage techniques don't always encourage a decrease in N_2O emissions; in certain cases, they even increase emissions, which are dependent on the properties of the soil and the local climate. Adopting zero tillage methods instead of tillage and soil-disturbing practices can cut CH_4 emissions by up to 21 and 25%(Tóth et al., 2018). Similarly, in the loamy soil of Southern Brazil, a reduction in CH_4 emission was noted with the implementation of no-tillage practice as opposed to the old tillage method. The soil quality can be ascertained by connecting the biological characteristics of soil, such as microbial activity and its functions, to the physical features. Out of all the functional potentials of agroforestry, carbon storage and sequestration are becoming increasingly recognized for their tendency to lower greenhouse gas emissions in the atmosphere, mitigate climate change, and slow down global warming by fixing atmospheric carbon into soil and vegetation. Poor farmers use these basic materials as a source of revenue as a means of boosting livelihood stability. The amount of biomass produced by AFs in rainfed, arid, and semiarid regions ranged from 2 to 10 tons/ha/year. Therefore, the choice to implement agroforestry and its management is never useless, and its usefulness in managing resources and mitigating climate change is remarkable. Agroforestry is practiced across the tropics due to its extensive adoption in agro-ecological regions and its diverse range of tree-crop combinations. This includes its attributes of economically profitable and sustainable land use systems.

AFs offer better potential for managing natural resources and utilizing them than other conventional agricultural techniques. They can also conserve biodiversity and produce more biomass. However, poor productive soil with limited availability of vital nutrients, acidic soils, and highly leached soils pose a challenge to subsistence farmers in the tropics, and these farmers typically lack access to mineral fertilizers for their farms. They also provide benefits such as improved soil fertility and quality, carbon input, organic matter addition, and nutrient status upkeep through litter inputs other than biodiversity conservation. Soils are the largest pools in the terrestrial ecosystem. They can provide diverse ecosystem services, such as acting as a fundamental substrate for plants and animals, storing billions of micro- and macroorganisms, fungi, protozoa, and earthworms, and storing vital elements that are crucial to plant growth and development as well as the maintenance of structure and functions of ecosystems. Effective nutrient cycling is preferable for agroforestry in single cropping systems. The provision of ecosystem services through soil is becoming widely acknowledged in terms of preserving soil fertility, enhancing soil structure, sustaining nutrient loads, supporting microorganisms, effectively cycling nutrients, storing, and sequestering carbon, and so forth. All of these services are critical for preserving soil health. Litterfall and other waste from trees in the AFs can contribute to the soil organic matter, which breaks down into simpler forms with the aid of microorganisms. This can be used by the tree crop system to sustain the health and overall productivity of the ecosystem. The disintegration of the subsurface root system can contribute vital nutrients and improve soil fertility; these nutrients serve as the main supply for the tree crop system(Kass et al., 1997).

Figure 3.3 Agroforestry

Source: https://en.wikipedia.org/wiki/Agroforestry#/media/File:Agroforesterie_(ma%C3%AFs_et_ch%C3%A2taigner)(2).jpg, *Creative Commons License.*

3.6 AGROFORESTRY

Agroforestry (Figure 3.3) has the potential to enhance soil biological systems, including earthworms, fungi, protozoa, and other soil-inhabiting organisms. These organisms are essential for the disintegration of organic matter as well as the proper distribution and translocation of that matter in plants, which is a sign of a healthy ecosystem structure and functions. For instance, earthworms, which are beneficial to farmers, are well known for their ability to break down and degrade coarse organic matter and distribute it properly from lower to upper soil layers, as well as in reverse, maintaining the condition of the soil mineral and organic layers. Furthermore, the breakdown of organic matter can provide carbon to the soil, and the amount of carbon that accumulates in soils can vary depending on the kinds of trees, their characteristics, the sorts of litterfall they produce, how quickly it breaks down, the species that break down organic residues, etc. In addition to adding additional carbon to soils, this input can help mitigate the current global warming phenomenon and play a fundamental role in sequestering and storing carbon. To increase soil fertility and nutritional status, AFs add litter and other waste to the agroforestry soils/platforms. This decomposition releases vital nutrients, which leads to close and efficient nutrient cycling and a healthy ecosystem and environment. In agricultural farms, integrating trees (wood/timber) with herbaceous crops promotes an effective close nutrient cycle, improves nutrient availability, delivery, and distribution in the root zone of crops, and reduces nutrient losses. While nitrogen cycling and its management represent a new paradigm that is closely associated with increased productivity and profitability, we are aware of the potential benefits that AFs can have for the environment and the economy. Nonetheless, AF methods ensure improved soil health by boosting fertility in a variety of soil-plant systems and supporting ecological processes. Similarly, biological components of soil under AFs and other forest systems are involved in the decomposition and decay of organic residues, which is crucial for managing fertility and nutrient cycling.

3.7 FIXING THE SOIL

Research has determined how soil microbial biomass, organic matter, and soil fertility in different AFs are related to key nutrients including carbon, nitrogen, and phosphorus. Important bacteria that live in soil play a minimal role in the cycling of nutrients as well as the breakdown of organic materials. While soil microbes can provide valuable insights into how soil health and quality are changing, different agroecosystems have different soil types and different management approaches require different environmental conditions. Thus, the population and functional activity of residing microorganisms, as well as their role in the breakdown and decomposition of organic matter and the release of nutrients and availability to plants, determine the productivity of AFs and the nutrients in the soil.

Leguminous plants have a great capacity to fix atmospheric nitrogen with the aid of Rhizobium bacteria located in root nodules, which has a favorable effect on AF productivity and soil improvement(Franche et al., 2009). This fixed nitrogen can be released through litterfall and its breakdown, along with root decay. Climate change is the main environmental burden that impacts the morphological, anatomical, physiological, and phonological changes that occur in all species, including the plant community, as well as a decline in productivity and health. Developmental initiatives, industrialization, and other man-made and natural activities release carbon dioxide, one of the atmospheric gases, along with other greenhouse gases. Forests and AFs contribute to soil storage of soil carbon as well as the absorption and fixation of atmospheric carbon and other gases into perennial trees and their constituent parts, including leaves, branches, and underground root systems. However, the greatest potential for land reclamation and CC reduction through the carbon sequestration process is seen in the integration of woody perennial trees with some pastures/grasses under silvopastoral systems.

Furthermore, the types of species, their absorption capacity, the types of agroforestry models, their potential for absorption, and the current climate all have a direct impact on the fixation and distribution of carbon. Agroforestry techniques, such as the production of biofuels and other fuels, are an excellent way to lessen reliance on fossil fuels, which are the important cause of greenhouse gas emissions into the environment. Including a fast-growing tree in the agricultural farming system is a sensible decision since it aids in the absorption of carbon and its retention in the woody portions of the tree as well as the system, which is based on the carbon farming system. AFs in humid tropical regions are estimated to maintain high carbon, which ranges from 0.3 to 15.2 MC per hectare year(Raj et al., 2019).

There is a correlation between improving livelihoods and tree-based farming techniques. In addition to ensuring food and environmental security, AFs guarantee income by supplying farmers with a variety of products, including fuelwood, timber, and non-timber plant products like edible fruits, medicinal plants, gum, resin, etc., which are essential to the survival of the rural poor. Agroforestry practices include gum exudation, apiculture, lac culture, and sericulture. These practices are carried out in a modern and scientific method and can provide farmers with additional income while also maintaining their socioeconomic status and promoting biodiversity conservation. Furthermore, the production of gum and resin under AFs increases farmer income in Ethiopia by 14 to 23 percent(Worku et al., 2014).

Farmers, planners, and legislators are becoming increasingly aware of AFs these days. As a result, location-specific models have received increased attention, and related research has been given priority. The last ten years have seen changes in agroforestry research and development, with a greater emphasis on the scientific and rigorous study of agroforestry models. To address challenges related to land use, socioeconomic concerns, managing soil health and quality, and addressing food and nutritional limits in the situation of climate change and extreme weather events, scientific research and development should be conducted. The research agenda should focus on improving soil fertility, health, and quality; tree-crop interaction; socioeconomic challenges ranging from small-scale to large-scale landscapes; and preserving food and environmental security with increasing attention. Other areas that require scientific research include C storage and sequestration, biodiversity conservation, wasteland development, and the rehabilitation of degraded soils. In light of the changing climate, policymakers, governments and non-governmental organizations, NGOs, organizations, etc., should collaborate to support the expansion of agroforestry regions. The many advantages of agroforestry and its ecosystem services, which support human life and preserve nature, should be known to the public and farmers. Therefore, there is a need to work toward improving the environment and ecosystem while bolstering national security. This includes the roles of AF role in maintaining FNS, resource conservation, soil health management, storage, and sequestration of carbon in both vegetation and soils, socioeconomic improvement of poor people, and provision of several important ecosystem services. Urban regions are linked to many issues since they continue to house a large portion of the global population, particularly in Asia, Latin America, and Africa. Many low-income urban people are the result of these socioeconomic issues, which also include high unemployment rates, poverty, substandard housing, and food insecurity. Due to their socioeconomic conditions, a large number of urban residents are compelled to cultivate and raise their own food to avoid extreme poverty and hunger. Furthermore, there are other obstacles to urban farming in underdeveloped nations, mainly being access to capital, water, and electricity. This will require them to make the most of all the opportunities that are within them. For instance, African and underdeveloped nations can use solar energy for vertical farming instead of relying on LED lighting systems.

3.8 OTHER OPTIONS

Small farms can be grown by entrepreneurs using low-tech techniques that require less funding, water, and light. For example, the Uganda population struggled to establish modern vertical farms due to a lack of funding and restricted access to land and water(Giller et al., 2021). Despite these challenges, urban farmers in Uganda have come up with creative ways to store vegetables, such as wood-designed crates. Modern vertical farms in Kenya have been constructed. All these designs were chosen to optimize water use and shield the vegetables from insects. In industrialized nations, urban farming produces food for the metropolitan areas but also applies advanced technology to sophisticated farming endeavors in the urban setting. Urban agriculture has benefited from the application of discoveries in areas such as water distribution, retention, protection, wetlands drainage, etc. In industrialized countries, urban agriculture is an unconventional practice that might involve community gardens. It involves networks of interested food-related groups or policymakers as part of a traditional process. In developing nations such as the United States, urban agriculture (Figure 3.4) is practiced by individuals who engage in food farming as a group activity, such as gardening or planting trees on public or private properties.

Figure 3.4 Urban agriculture

Source: https://en.wikipedia.org/wiki/Urban_agriculture#/media/File:New_crops-Chicago_urban_farm.jpg, *Creative Commons License.*

The Netherlands produces high-quality, pesticide-free vegetables by using large commercial indoor farms. These indoor farms use artificial climate control systems and advanced lighting systems to cultivate certain crops in structures that need less soil and water to grow. It is a more advanced scientific method that involves farms running in a closed-growth environment. Because it helps regulate plant evaporation, these farms typically use less water than conventional farms. Furthermore, the majority of vertical farms, or aeroponics or hydroponic systems, demand that plants be grown in nutrient solutions with access to air and water rather than always utilizing soil as a medium. It will satisfy both the growing need for high-quality, pesticide-free food and the needs of the growing urban population. The food supply for urban dwellers is probably at risk if there is persistent growth in the population without corresponding improvements to farming and food supply techniques. This may raise the likelihood of food shortages in urban areas across the world. In recent decades, there has been an unintentional marginalization of urban agriculture in cities due to the necessity for a conveniently available urban food supply.

3.9 SOLUTIONS

Urban agricultural activities serve multiple purposes. Firstly, they can reduce poverty and hunger among urban residents. Secondly, by engaging in agricultural activities, income can be generated through product sales. Russian urban farming and gardening practices bring in enormous amounts of money.

Urban farms make money by making a profit and selling food. Small farms with less than $250,000 total sales can be classified as commercial farms. Urban agriculture should be promoted since it improves the economic status of urban regions and lowers unemployment by offering opportunities that create money. The advantages include developing jobs and encouraging fruit and vegetable growth, easy access to wholesale organic food, enhanced nutrition, the reduction of climate change, photosynthesis, and many more advantages. Urban agriculture generally contributes significantly to sustainable development through the creation of open green spaces, the diversification of urban habitats and, therefore, biodiversity in cities, the reduction of noise and pollution, and the mitigation of climate change. In general, urban agriculture has greatly benefited society, particularly in terms of food and vegetable supply. But in cities, a few issues prevent it from reaching its full potential. These issues include the seasonality as impacted by CC, the lack of usable landscapes, high start-up costs, and traditional gardening techniques.

The lack of awareness and attention given to agroforestry promotion in rural regions is likely a result of its incapacity to yield substantial financial rewards. On the other hand, through community-based natural resource management (CBNRM) initiatives, local communities have actively participated in the management and conservation of wildlife. However, even though many rural families rely on these resources for survival, CBNRM initiatives have not received enough promotion. Nearly 200 million indigenous tribes rely solely on forests, and over 1 billion people are thought to depend on the many forest resources for their survival globally. Furthermore, 350 million people who reside close to dense forests rely on them for both income creation and survival(Razafindratsima et al., 2021). Additionally, it has been estimated that in developing nations, the majority of the income of the rural people comes from locally available environmental resources that can support them in times of crisis or food shortage. When top soils, trees, and other vegetation are removed for mining purposes, the landscape is destroyed, a significant amount of vegetation is lost, and the lives of people, animals, and plants are put in risk. The resources found in forests are under tremendous strain due to the rising demand for natural resources. As a result, mining operations have a detrimental environmental impact on trees and other vegetation and generally worsen CC problems. Thus, more agroforestry education and awareness efforts are urgently needed to emphasize the significant advantages of agroforestry promotion or re-afforestation initiatives in rural regions. Changes in rainfall, temperature, moisture content, evaporation levels, and soil temperature at different depths are only a few of the ways that climate and weather conditions affect and restrict all-season farming. Aggressive agriculture will also be a part of all-season farming. One of its main benefits is that crop yield will rise. As a result, farmers can easily keep an eye on the land, safeguard their cattle, and produce a lot of inexpensive food, fruits, vegetables, and poultry.

3.10 NEW QUANTIFYING APPROACHES

The quantification of aboveground and belowground biomass is typically the main focus of measurements of the CS of Afs(Dhyani et al., 2020). Indirect remote sensing methods, modeling, and on-site direct biomass measurement can all be used to complete these observations. The most common method for directly measuring aboveground biomass on site is to either fully or partially harvest the tree before measuring it. To put it simply, 50% of the total biomass is commonly regarded as the biomass carbon content. Although it takes a long time and costs a lot of money, the total destructive method of biomass measurement is accurate and is typically used to confirm the findings of other methods. This approach divides the harvested trees into many sections, such as barks, leaves, branches, and fruits. Each part is measured independently and added together to generate the total aboveground biomass. Regression models can be used to frame equations by identifying independent and dependent variables in the data. Tree growth characteristics such as height and diameter are typically employed as independent variables. Soil carbon and root biomass are included in the estimate of belowground biomass carbon. Measuring belowground biomass, particularly for roots, is a labor-intensive process that calls for consistent estimation. One method for measuring root biomass is to fully excavate the roots or use spatially scattered cores or pits. Therefore, assessing the root-to-shoot ratio is the simplest method for calculating the root biomass. The carbon content of the root biomass can also be calculated using the factor or direct estimation in the laboratory analysis, just like the aboveground biomass. A conventional approach can be used in a laboratory to estimate the carbon content of soil. Usually stated as a percentage or as g/kg, the amount of carbon content in soil can then be extrapolated to be expressed as mg ha^{-1}.

The United Nations Framework Convention on Climate Change, or UNFCCC, has underscored the significance of sinks in maintaining the CO_2 equilibrium and the capacity of various land use systems to store carbon. The tree-based land use system is shown to have the highest and most promising carbon sink potentiality among the various land use systems. The REDD programs of UNFCCC have concentrated on lowering emissions, mostly caused by deforestation and forest degradation. In this context, agroforestry is beginning to gain popularity across the globe due to its capacity to be more flexible and resilient to the effects of climate change, and diversify the revenue streams. In terms of agroforestry, carbon sequestration (CS) refers to the process of photosynthesis that removes and stores atmospheric carbon in the biomass of trees over an extended period. As soil organic carbon (SOC) increases under Afs, a significant amount of carbon is stored in the soil ecosystem. Agroforestry tends to store and collect more atmospheric carbon in the biomass of plants and soil because it is a more complex and diversified system than typical monocropping.

Chapter 4

CLIMATE CHANGE AND PUBLIC HEALTH

Contents

4.1 INTRODUCTION

Either natural change or human activity can contribute to climate change, but it all depends on how well solar shortwave radiation and outgoing infrared longwave radiation are balanced. The atmosphere similarly influences this balance to how sunlight enters a greenhouse or a car windshield, trapping heated air below and preventing it from ascending and combining with cooler air above. Over the last decades, concentrations of GHGs such as CO_2, methane, nitrous oxide, and others have significantly increased(Schulze et al., 2009). Between 2005 and 2011, the increase was as high as 43 percent. For instance, since the start of the Industrial Era, the quantity of CO_2 in the atmosphere has significantly increased. These three GHGs are present in the atmosphere at levels higher than those that have been observed during the previous 800,000 years. Furthermore, these three GHGs have increased faster over the last century than they have at any other point in the previous 22,000 years. Increased average surface temperatures result from an environment that retains more heat due to increased GHG concentrations. The average surface temperature rose by 0.85°C (1.53°F) between 1880 and 2012(Anting et al., 2017). Reduced biodiversity in ecosystems due to changes in chemicals and aeroallergens in ambient air, diminishing glaciers and polar ice caps, and other environmental phenomena may also be linked to climate change. According to the IPCC, global warming since the 1950s has been unprecedented. It has linked human activities to this warming with 95% certainty. According to IPCC projections, the average temperature of the surface will rise by 1.0° to 3.7°C (1.8° to 6.7°F) by 2081–2100(Mostafa et al., 2019). The frequency of daily temperatures over 38°C (100°F) is predicted to rise significantly in the United States; temperatures that currently occur once every 20 years may suddenly happen every two years. There is a projected increase in the duration, severity, and frequency of extreme heat episodes. In certain regions, extended drought-related high temperatures lead to dry conditions, which raise the possibility of wildfires.

4.2 EL NINO

El Nino-related precipitation variability is expected to increase globally, while mean precipitation is expected to decline in many arid regions. It is anticipated that precipitation will grow more severe but less frequent in the United States. It is predicted that there will be more heavy rainfall episodes because warmer air retains more water vapor. Except for Hawaii, the United States has seen above-average precipitation during periods of intense rainfall since 1991. In addition to mental health effects, heavy rainfall events are most likely to result in flooding, which can harm individuals directly and indirectly by contaminating drinking water with sewage. The frequency of flooding in rivers has been rising. In the years between 1980 and 2009, floods killed over 500,000 individuals and negatively impacted at least 2.8 billion people(Done, 2012). Globally, and mainly in Asia, Africa, Central America, and South America, an increase in floods is predicted. Because of the combination of rising temperatures and falling precipitation, climate change is predicted to cause a rise in the frequency of wildfires.

Figure 4.1 Wildfires

Source: https://en.wikipedia.org/wiki/Wildfire#/media/File:Prescribed_burn_in_a_Pinus_nigra_stand_in_Portugal.JPG, *Creative Commons License.*

Wildfires (Figure 4.1) increase air pollution and have a negative impact on health both directly and indirectly. Air pollution from forest fires is thought to be the cause of between 260,000 and 600,000 premature deaths worldwide each year. As human populations spread into flood plains and vulnerable coastal areas, the effects of floods and coastal storm surges are likely to worsen. There is a greater chance of landslides in regions where there has been deforestation due to high rains. For instance, 11,000 people died in Central America in 1998 as a result of Hurricane Mitch, many of which were brought on by mudslides from deforestation(Haigh et al., 2004). In 2005, thousands of people were displaced, many more were injured, and over 1,800 people died as a result of Hurricane Katrina and the subsequent flooding. The loss of coastal wetlands, which provided less storm surge protection, most likely made the impacts more severe. Labor productivity will often decline at work, which will have significant financial ramifications for communities, enterprises, and families.

4.3 CLIMATE CHANGE

In many parts of the world, rising temperatures or heat waves are the most obvious result of climate change. Additionally, the effect of rising greenhouse gas (GHG) concentrations in the atmosphere is most frequently modeled. The primary method for measuring ambient environmental heat is air temperature. However, heat is a function of air temperature, humidity, air movement (wind speed), and radiation. Heat exposure is a health risk associated with an environment that has not received enough attention. It is caused by a combination of air temperature, humidity, air movement, and radiation. Both clothes and metabolic rate, a product of muscular movement with physical activity, have an impact on an individual's susceptibility to heat stress. The physiological balance of body temperature and hydration is threatened by humid ambient conditions with air temperatures above 35°C (95°F) unless fluid lost through excessive perspiration is restored with drinking water. The development of clinical diseases and decreased human performance are particularly common in youngsters, the elderly, and those engaged in physically demanding occupations. More than 40 heat stress indices can be used to characterize heat exposure and stress. The popular ones include the Universal Thermal Climate Index (UTCI), the heat index (HI), the corrected effective temperature (CET), and the wet bulb globe temperature (WBGT). There are benefits and drawbacks to each index. WBGT can be interpreted in relation to physical activity levels in daily life or at work. WBGT is frequently utilized in national standards and is endorsed as the recommended heat index for international standards. It can be understood in terms of clothes and metabolic rate. In most tropical countries, during the hottest seasons, daytime and overnight heat levels exceed what is acceptable for daily living or for achieving the necessary productivity at work when efficient cooling systems are not available. Even in moderately intense jobs, human performance and labor capacity are limited by a WBGT of 29°C (84°F).

4.4 TEMPERATURE-WORK AND LEISURE

Increasing temperatures are causing an increase in heat-related problems in the southwest of the United States, Mexico, Cuba, Southeast Asian countries, and other regions. WBGT levels are already high enough in several countries to restrict both indoor and outdoor work. For instance, in Singapore, from roughly 10 in 1980 to approximately 70 in 2010, the number of days at or above an indoor WBGT of 29°C (84°F) has increased annually, indicating a notable increase in the length of the extremely hot season each year(Ioannou et al., 2022). Not only does work suffer from these high temperatures, but many other daily activities do as well. Given the current trends, by 2050, the situation in these kinds of locations will be worsen. In industries with high heat exposure, people lose around an hour of work time per day when the maximum temperature rises above 35°C (95°F). Low-risk industries," such as those with predominantly indoor employment with air conditioning or other cooling systems, were less severely impacted than high-risk industries, such as those with jobs involving hard labor outside. There was less time spent outside and more time spent inside during the warmer months. Additionally, unemployed individuals spend more time indoors than outdoors, which may indicate a shift in their leisure activities. Recreational and outdoor work are diminished when climate change results in longer and more intense hot seasons.

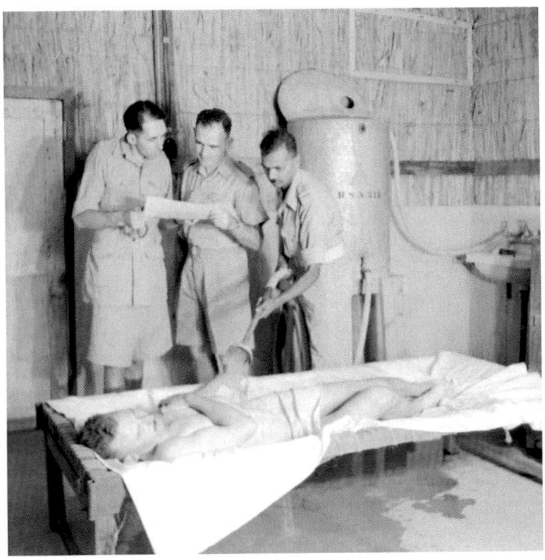

Figure 4.2 Heat stroke

Source: https://en.wikipedia.org/wiki/Heat_stroke#/media/File:The_British_Army_in_the_Middle_East_1943_E26027.jpg, *Creative Commons License.*

It was hypothesized that the projected decrease in time use related to productivity losses at work caused by heat. More research is needed on the effects of workplace heat on productivity. Many case studies that have been published from various countries show instances of heat-related diseases that can occur both indoors and outside while working. Circumstantial evidence and observations have been gathered from low-income tropical countries in factories devoid of effective cooling equipment, such as air conditioning, across a variety of industries, including the production of inexpensive clothing, shoes, and other consumer products intended for export to wealthy nations. Studies have shown that heat poses serious risks to workers at all times of excessive heat, not just during heat waves. Many extensive studies conducted in Thailand have demonstrated that jobs with high heat exposure are associated with lower self-reported overall health status, higher renal disease prevalence, and higher histories of occupational injuries. Additionally, heat-related deaths have been documented in Japanese workplaces, among men of working age during the 2003 heat wave, and among immigrant laborers from Nepal and India who were constructing a soccer stadium in Doha, Qatar.

4.5 HEAT STROKE

Because their ability to sweat and sense heat stress both decline with age, older persons are particularly vulnerable to heat exposure (Figure 4.2). Studies conducted on senior citizens in Japan found that they become less active in the heat. Anthropological research conducted in India (1971), revealed how the residents prevented heat stroke by modifying the amount of hours worked and the intensity of activity to suit the local climate. While people who work and live in hot environments can adjust their behavior in these ways, daily activities that are restricted due to heat diminish the quality of life in these environments. The ideal body core temperature is maintained at or around 37°C (98.6°F). There is a heat exchange between the body and its immediate surroundings. Heat energy is lost from the body when it is warmer than the surrounding air and is added when it is colder. Eighty percent of the energy utilized by muscles is converted to excess heat inside the body, which is difficult to release in hot ambient settings. This heat is produced by both basal metabolism and additional metabolic activity. Heat is released from the body by the evaporation of sweat. The humidity of the surrounding air, however, affects the rate of evaporation. Evaporation slows down as humidity rises and stops completely at 100 percent relative humidity. However, if all of the liquid lost through perspiration is not replenished by drinking water, prolonged sweating may cause severe dehydration. Blood flow to the skin needs to rise for the sweat evaporation mechanism to work properly. Increasing cardiac output and enlarging skin capillaries are two ways that one can improve their capacity to remove body heat during physiological acclimation to high temperatures. Heat stroke is more common in those who cannot quickly raise their cardiac output, such as many elderly persons. Heat effects on human health and performance are influenced by a number of elements. These include the metabolic rate resulting from working, walking, or cycling as an active mode of transportation, playing sports, and the impact of clothes. Clothing that is insulated helps lessen the internal and external heat inputs and losses. However, clothes can hinder evaporation when a person is sweating, which might restrict cooling. For instance, workers who must wear protective clothing while working in hot surroundings will not benefit fully from sweat evaporation. Thus, hot conditions with air exceeding 37°C (98.6°F) pose a hazard to physiological equilibrium, particularly for workers wearing protective clothing or other standard protective gear.

The negative effects of heat are partly attributed to cultural and economic reasons(Banwell et al., 2012). Heat-related issues are lessened if employees don't work during the hottest part of the day. However, this results in fewer hours being available for work. Moving labor and other demanding daily activities to later hours of the day or cooler seasons of the year might also help minimize heat-related issues. These modifications, nevertheless, might make labor less productive. For instance, women are expected to dress traditionally in some places, even if their jobs require them to wear distinct protection gear. Therefore, jobs that generate significant internal heat production might become very dangerous for women in these fields. There is little access to automated machinery at work, particularly in low-income countries. Thus, the primary energy source for production, especially in low-income nations, comes from manual labor. Future predictions are dependent on local and national conditions because heat patterns vary between and within countries. By 2046–2065, for instance, there will likely be a significant increase in the number of summer days in New York with a maximum temperature higher than 32.2°C (90°F). Between 1975 and 2000, there was a significant rise in summertime heat exposure in the United States and the Caribbean. In certain places, the daily maximum WBGT rose from 27°C (81°F) to 30°C (86°F)(Budd, 2008). There were almost 50% fewer afternoons available for doing physically demanding tasks. WBGTs would significantly decrease productivity and raise the risk of heat-related diseases if they were to rise by an additional 3°C (5.4°F). Heat has similar effects on human performance and day-to-day living in many other parts of the world, including much of Latin America, the Pacific islands, Australia, China, India, Southeast Asia, and much of Africa. There will be a significant rise in the number of days in Perth, Australia, where heat stress will make it difficult for people to engage in outdoor physical activity.

4.6 WORKPLACE CONDITIONS

The only preventative adaptation strategy taken into account in estimations is decreasing heavy labor outside. Nevertheless, the anticipated economic impact of decreased productivity at work is probably considerably higher than the other effects of climate change, like negative health effects. While most Western European countries experience little to no loss, low- and middle-income countries are most badly impacted. The two most important preventive steps to address the effects of heat stress are having heat-warning systems and being prepared. However, creating such solutions is difficult, and exchanging knowledge about local conditions and heat-related hazards can help. It is particularly crucial to have access to inexpensive data loggers to monitor the humidity and temperature in work environments. Increasing the use of conventional air conditioning alone will not address the issue of human heat exposure in homes, workplaces, or other locations. This is because conventional air conditioning typically uses electricity generated from non-renewable energy sources, increasing GHG emissions, which worsens the fundamental cause of climate change. Furthermore, the urban heat island effect is amplified when air conditioning systems are used in metropolitan settings. In Bangkok, for instance, an increase in external heat of 1°C (1.8°F) was linked to a rise in electrical demand of almost 2,000 megawatts(Worawiwat et al., 2021). Furthermore, air conditioning is unaffordable for hundreds of millions of people living and working in low-income countries. It is also important to remember that using air conditioning on a daily basis lessens the degree of heat acclimation and raises heat strain when a person exits the air-conditioned area. In certain situations, alternative techniques for cooling indoor areas could be useful in lowering heat stress. The surface evaporation technique is a simple method that lowers air temperature and increases evaporation by hanging big, wet blankets over a workstation. This method acts to enhance relative humidity, which may increase heat exposure, even if it can be useful in interior rooms with low relative humidity and enough air movement. Fans pointed toward certain employees or groups produces a feeling of cooler air movement is produced, and the air flow promotes sweat evaporation. Fans, however, have less cooling power when the outside temperature is higher than 35°C (95°F), and there is a lot of humidity.

4.7 NOISE AND OTHER STRESS

Furthermore, fans can expose workers to dangerous levels of noise in many indoor workspaces(Hernandez Bennetts et al., 2019). Constructing buildings close to trees or other structures that provide shade is another way to lessen the amount of heat exposure outside. Another practical way to increase shade is through urban design, which is especially helpful in tropical and subtropical nations that are quickly urbanizing. It is possible to design colder buildings and require less air conditioning by using architectural techniques to prevent interior heat loads. Heat traps for sunlight are created in slum areas by black-painted or black-colored asbestos cement or corrugated iron roof sheets, which transfer heat to the interior space below. Slum dwellers consequently have a higher risk of heat stress. In any metropolitan region, there may be an issue with the additional heat absorption from solar heat radiation from black or extremely dark-colored materials. Similar to how modern building design might incorporate air ducts that allow cooler air from below or lower levels to rise to higher levels in a building, traditional building techniques in tropical hot places have understood that wind movement can reduce heat. All architectural approaches in hot environments must be sustainable for a building life cycle. Even if present solutions to adapt to climate change are deployed, millions more people will find themselves living and working in areas of intense heat stress for extended times of the year as average temperatures rise and spells of extreme heat get longer. Thus, it is obvious that taking action to slow down climate change is essential. Heat-warning systems, better public readiness based on weather forecasts, enhanced formulation and enforcement of health and safety standards, and training programs for managers, supervisors, and employees are important components of initiatives to lessen the heat impact of climate change. Adaptive solutions will be necessary in industries that employ high-risk personnel, such as construction workers, immigrant agricultural workers, and those who conduct rigorous work without efficient cooling equipment.

Figure 4.3 Air conditioning systems

Source: https://en.wikipedia.org/wiki/Heating,_ventilation,_and_air_conditioning#/media/File:Rooftop_Packaged_Units.JPG, *Creative Commons License.*

4.8 AIR CONDITIONING SYSTEMS

With regard to heat exposure and climate change, several preventive actions can offer certain advantages. Some air conditioning systems, (Figure 4.3) for instance, use solar heat radiation as the energy source for the cooling process. As a result, these systems offer carbon-neutral cooling, which cools indoor spaces without increasing emissions in the way that installing air conditioning would have otherwise done. The report from the Intergovernmental Panel on Climate Change or IPCC describes additional measures for mitigating and adapting to the heat-related effects of climate change that also have positive health effects. Climate change is projected to result in an increase in morbidity and mortality from heat stress. We must understand pathophysiology, sensitive populations, current heat wave situation, epidemiological research, future projections, and preventive actions. Research on the health effects of exposure to high ambient temperatures typically employs time-series models, such as Poisson regression in their epidemiological investigations. Instead of focusing only on the heatwave(s), some studies evaluate the effects on health in a specific geographic area, frequently throughout the course of an entire warm/hot season. This method takes the long-term effects of heat exposure on public health into account and enables a more comprehensive representation of widespread heat exposure. Study design can compensate for confounders like air pollution, and the effect of temperature persisted even after accounting for air pollution. Researchers must characterize a specific heat wave to concentrate just on it. This can be done by providing daily mean, maximum, and minimum temperatures over a period of days, along with percentiles.

4.9 EFFECTS OF HEAT WAVES

The health impacts of one heat wave, however, might not be indicative of the health implications of subsequent heat waves; these depend on the area and other risk factors both within and between geographic regions. For instance, a heat wave in London can have quite different health impacts than one in Athens. Nevertheless, in addition to heat-related disorders, there is proof linking heat waves to the beginning or worsening of every ailment.

The body may gradually adjust to rising temperatures by releasing more heat through various systems, but in certain situations, as during a heat wave, this may not be possible in a short amount of time. Vasodilation lowers blood pressure and raises serum cholesterol and blood viscosity during hot weather, both of which can surge the risk of ischemic stroke or myocardial infarction(Widlansky et al., 2007). Blood coagulation changes may raise the risk of thrombosis. In susceptible individuals, a higher threshold for sweat may also have detrimental effects on health. Pregnant women may not be able to thermoregulate effectively, similar to other populations susceptible to the effects of extreme heat. This may be because of heat-related dehydration, which can reduce uterine blood flow and increase pituitary secretion of oxytocin and antidiuretic hormone to induce labor. A portion of the diseases and fatalities linked to climate change are indirectly induced. For instance, warmer weather can exacerbate several respiratory conditions by increasing exposure to outdoor air contaminants like ozone. Asthma and allergies (particularly in children), respiratory symptoms and reduced lung function, missed school days due to respiratory diseases, ER visits, hospitalizations, and fatalities are some of these conditions and their associated effects. Low-income communities account for a disproportionate share of mortality attributable to ozone depletion. Most of the United States has ground-level ozone levels that are higher than federal and state regulations, particularly in densely populated areas. Future projections indicate that these levels will rise. During heat waves, there are two types of risk factors for heat-associated disorders: extrinsic and intrinsic. The effects of heat waves on health differ depending on factors such as geography, air pollution levels, whether air conditioning is used or not, education, socioeconomic status, age, gender, and race/ethnicity.

Figure 4.4 Outdoor workers

Source: https://en.wikipedia.org/wiki/Live-line_working#/media/File:Lineman_changing_transformer.jpg, *Creative Commons License.*

Mentally ill people and those with certain pre-existing medical diseases are also at risk. Individuals with diabetes, depression, cardiovascular and cerebrovascular disorders are more susceptible to the negative effects of heat. People who suffer from diseases that affect their awareness, movement, and conduct are also more vulnerable. The risk of heat-related diseases and deaths has also been demonstrated to increase with the use of certain medications that may impair sweating, such as diuretics, anticholinergic agents, and tranquilizers. The risk of morbidity and death from heat-related diseases is further increased in individuals over 50, as well as in young children and babies, who have less ability to thermoregulate (Faurie et al., 2022). However, even healthy individuals may die from a heat-related disease if their exposure to heat is extreme enough. Outdoor workers (Figure 4.4) are more susceptible to heat-related diseases and fatalities when they work in extremely hot environments. Furthermore, excessive heat can negatively impact many daily activities and work performance. Australia will not be able to engage in outdoor activities for

33 to 45 days a year in 2070, as opposed to the current 4 to 6 days(Maloney & Forbes, 2011). Because they are more likely than women to be active in hot weather, men are at higher risk of dying from heat stroke, which is particularly related to their exposures. Due to the urban heat island effect, people who live in urban areas may be vulnerable to the negative effects of heat waves. Cities may see an increase in average surface temperatures of roughly 0.27°C (0.49°F) due to the urban heat island effect. Buildings, industrial and human activity, and other factors cause cities to produce and hold onto heat, which is what is known as the urban heat island effect.

Nighttime temperatures are raised by urban heat islands because they radiate heat at night. Cities generally have fewer trees, which adds to the urban heat island effect. The average temperature in California is expected to rise by up to 2.8°C (5°F) by the 2030s and up to 5.6°C (10°F) by 2100 or earlier(Messner et al., 2011). In Southern California, a drier climate with proportionately less snow and more rain is anticipated. This will result in a decline in the quantity and quality of water, a reduction in food output, and a rise in waterborne diseases. Increased temperatures, droughts, and early snowmelt will cause the soil to dry and raise the risk of wildfires. More harmful than the particulate matter emitted by cars, wildfire smoke can aggravate chronic lung and heart conditions, such as asthma, and irritate the eyes and respiratory system. In California, there were around 2,780 wildfires in the summer of 2008, which claimed 23 lives. In addition, there will probably be more wildfires as extreme temperatures that used to only occur once every century is now predicted to happen every year. Based on mean temperatures between 1961 and 1990 as a baseline, elevated emissions, and growing population, studies projected that heat-related mortality in California would result in 2,100–4,300 premature deaths by 2025 and 6,700–11,300 deaths by 2050(Ostro et al., 2011). The number of heat-related deaths in California that occur in the future is comparable to those of deaths from other key causes. It is predicted that the number of fatalities from heat-related causes would be roughly 3,700 per hot/summer season between 2048 and 2052, the number of deaths from cardiovascular disease will be roughly 3,500, and the number of deaths from all causes combined would be between 21,000 and 27,000 per hot season. Long-term environmental programs and short-term behavioral changes for individuals and communities are two ways to reduce heat-related diseases. During heat waves, effective short-term strategies include checking on elderly individuals, particularly those living alone; disseminating information about cooling centers; and, if necessary, giving transportation to cooling centers. The United States has several areas that are not prepared for heat waves. Several communities have created heat-response strategies in response to extreme heat waves to stop heat-related diseases and fatalities. Because the exposures and risk features of susceptible populations vary by geographic location, including socioeconomic situation and demographic considerations, heat-response programs must be tailored specifically to a city or region. Determining the moment at which heat waves become sufficiently risky to human health to require public health interventions is one of the challenges in creating heat-response plans. Too few heat advisories offer intolerable health hazards, and too many cause the public to disregard them as major emergencies. To ascertain whether preventive measures are effective, researchers must create new instruments, including computer programs that operate on the Web. Web-based computer programs that are currently in use include data on climate change variability to facilitate surveillance and offer graphical representations of projected changes in climate change within particular geographic areas throughout the world. Depending on the severity and situation, management options may include taking the patient's core body temperature, applying ice packs or chemical cold packs, hydration, cold-water therapy, passive cooling techniques like moving the affected person to a shaded area, and ice towels. Serious heat-related diseases/disorders, like heat stroke, necessitate hospital stays and frequently long-term care. While some areas have not developed heat-response strategies, others have made a good start in doing so. While many European countries have learned from the heat waves that occurred there in 2003 and 2010, others are still not prepared. For instance, out of the 51 countries in the WHO European Region, only 18 (35%) have created heat-response strategies. Significant shortcomings in the policies of some nations include concentrating on a single vulnerable segment, such as the elderly, rather than providing care for all high-risk groups. It may be challenging to identify those who are most susceptible to heat before a heat wave since they frequently are unaware that they are at high risk or choose not to be classified as such.

4.10 GROUND-LEVEL REALITY

Another important component in reducing the risk of heat-related diseases and fatalities is promptness. Plans for responding to heat waves should prioritize creating coordinated actions for populations at high risk as well as putting in place long-term monitoring and assessment. Heat-response plans have been created by certain Canadian cities. For instance, Montreal's heat-response plan was improved to enhance preventive efforts centered on informing people about weather warnings, and advisories to respond either before or during a heat wave after the 2010 heat wave, which caused approximately 300 deaths. The "Cool Toronto" initiative was created in response to the 1999 heat wave in Toronto(Smoyer-Tomic & Rainham, 2001). It focuses on realistic preventive actions for vulnerable individuals, like checking on elderly persons and giving transportation to cooling centers. Middle-aged farmworkers in the Central

Valley were among the vulnerable groups most severely affected by heat-related ailments during the California 2006 heat wave. Therefore, preventative measures are required to accommodate the unique needs and beliefs of farmworkers. Some misled beliefs held by farmworkers include the following: drinking water should be chosen based on location rather than whether it has been adequately treated; cooling treatments should be avoided after heat exposure; increased sweating may aid in weight loss; and highly caffeinated energy drinks can boost work efficiency. Employers frequently fail to give farmworkers proper clothing, frequent breaks, and enough water and shade. It is imperative that during heat waves or excessive outside temperatures, farmworkers and their employers understand the significance of limiting heat exposure using the most practical techniques. Improving working conditions can be facilitated by providing farmworkers and their employers with health education and promotion. Pregnant farm workers may be generally aware of the health concerns associated with heat exposure, but they are unaware of the risks associated with pregnancy and frequently have little control over their working environment. Another vulnerable subgroup includes athletes and others during heat waves, particularly those who work out outside. Athletes in high school and college are vulnerable to the negative effects of heat, but on hot days, they are frequently not sufficiently covered. Several California state authorities have produced guidelines for preventing heat-related diseases. By implementing these suggestions, local health authorities will be able to develop heat-warning systems that will help avert heat-related diseases and fatalities. These suggestions include assessing and implementing cooling techniques for both indoor and outdoor settings, creating an urban heat island index, strengthening the health care system capacity to withstand extreme heat, enhancing the promptness of heat-related disorder surveillance, assessing occupational standards, educating employers and employees who work outside about preventing heat-related diseases, identifying vulnerable populations for heat-related diseases and deaths, and filling in important research gaps.

4.11 OTHER IMPORTANT FACTORS

The factors of exposure, sensitivity, adaptation capability, and population density all affect susceptibility to climate change. Heat waves will have negative consequences on people in most countries, but they will have a higher impact on the poor and those from low-income countries because they will be exposed to more heat, both at home and at work, and they will have less access to air conditioning and medical care. People who are acclimated to heat exposure undergo biological acclimatization. Warmer areas typically have higher threshold temperatures for heat-related diseases. Cities in temperate zones require lower temperature rises to produce heat-related diseases than cities in tropical zones. Less natural adaptation is feasible with significant temperature increases; more preventive actions are required but frequently not taken. For low-income groups that are particularly susceptible, health education is crucial. Most people think that over the past 40 years, developed countries have been successful in improving the quality of their air. On the other hand, excessively high air pollution was caused by a multi-day temperature inversion across Western Europe in March 2014. Many children and adults have poor quality of life due to asthma and allergies, which also cause a large amount of premature mortality. Geographically and in terms of age group, the reported prevalence of childhood asthma varies widely; it ranges from less than 1 percent in Tibet to over 30 percent in some areas of New Zealand. Adults between 18 and 45 years are expected to have an 8.6% prevalence of wheezing and a 4.3 percent prevalence of asthma(Borna et al., 2019). Allergies to dietary and environmental factors are becoming more prevalent worldwide.

4.12 POLLEN AND OZONE DYNAMICS

Pollen levels are only regularly measured in a small number of cities, and measuring them needs skilled experts who have received the necessary training and certification. Since pollen is commonly measured in universities and private clinics, data are frequently not accessible to the general public. Consequently, the number of epidemiological studies on the effects of pollen is quite low. However, research has looked into how aeroallergens like ragweed, grass, and tree pollens affect asthma and S conditions like conjunctivitis, rhinosinusitis, and allergic rhinitis. Interpretation is challenging due to the fact that research on the health effects of aeroallergens varies depending on the type of disorder, disease prevalence, dominant taxa, aeroallergen mixture, pollen type, total pollen counts, statistical methods used, spatial representativeness of measured aeroallergens, duration of study period, and size of population studied.

As a result, a correlation that one city may find to exist between a certain type of pollen (Figure 4.5) or other aeroallergen and a respiratory or allergy condition may not hold true in another. Furthermore, the specific effects of climate change on a given type of pollen, such as its peak level and the length of its season, differ between and within countries. Due in part to regional variations in climate, each location has a very different pollen season. Many varieties of grass and tree pollen are produced in greater quantities during the pollen season due to higher temperatures and increased precipitation in the months preceding the pollen season. There is evidence that certain plant species in Europe have produced more pollen in the last 30 years, which is consistent with the warming over that time. The onset of a pollen season is also correlated with higher temperatures. In certain regions of Europe, Asia, and the US, certain plant species are already experiencing the pollen season earlier in the year, which is consistent with the recent rise in temperatures.

Figure 4.5 Pollen

Source: https://en.wikipedia.org/wiki/Pollen#/media/File:Pine_releasing_pollen_into_the_wind_in_Tuntorp_4.jpg, *Creative Commons License.*

Many studies have forecast future changes in ozone under different circumstances. While detailed results vary among studies, most suggest that climate change alone, even without changes in emissions of ozone precursors (VOCs and nitrogen dioxide), could lead to increased ozone episodes in polluted areas and decreased ozone episodes in rural areas, which are less influenced by emissions. Studies indicate that, depending on location and models, climate change may result in either an increase or decrease in $PM_{2.5}$, although $PM_{2.5}$ has received less attention than ozone. Models that predict reduced air pollution in industrialized countries due to better policies in the future typically show lower anticipated concentrations notwithstanding climate change; in other words, reductions in pollution emissions can counteract the effects of climate change. Projections of air pollution and climate change and their effects in developing countries, where pollution is generally rising and air quality monitoring and control are typically insufficient, have received far less attention. In the last seventy years, the number of people on Earth has tripled, and this, combined with growing needs for food, shelter, and fuel for cooking, heating, and transportation, has permanently altered the climate and ecology, allowing vector-borne infections to proliferate. Furthermore, a large number of people have relocated from rural to urban regions, which has increased the concentration of people living close to vital rivers and beaches and connected to them. When taken as a whole, these human-caused alterations have changed the atmosphere and climate, resulting in rising sea levels, intensifying storms, altered rainfall patterns, and overall warming, particularly in northern latitudes. Within urban centers, impermeable and reflecting surfaces have altered precipitation runoff and temperature patterns, creating urban heat islands. Vector-borne infectious diseases are particularly susceptible to climate change because the vectors are poikilotherms, or having body temperatures that are close to the surrounding air. Climatc and ccosystcm dynamics play a major role in determining host and vector abundance as well as seasonality(Altizer et al., 2006).

4.13 CHALLENGES IN MANAGEMENT

Management programs are frequently faced with the decision of whether to carry out preventive or reactive intervention when eradication is determined to be unachievable. Owing to the duration needed to implement extensive vaccine or vector control initiatives, public health authorities are forced to base their decisions on predictive models, historical climate data, and frequently scant monitoring information. The necessity of the costly intervention is questioned if it is successful and there are no cases. A changing environment that facilitates transmission, rapid connecting populations, and an increase in the number of people living in large cities mean that vector-borne diseases will continue to be a major global problem for public health and wildlife.

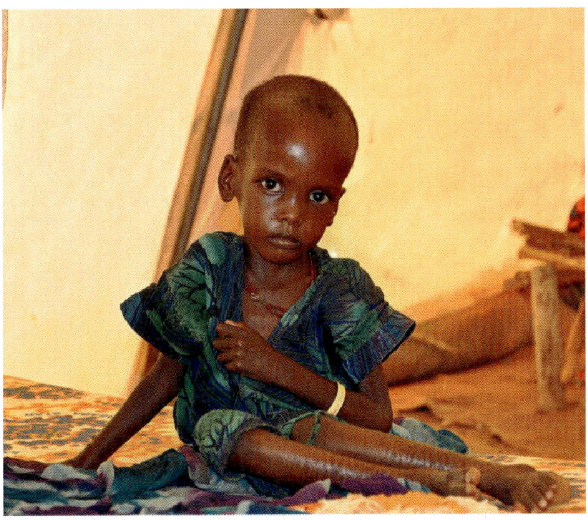

Figure 4.6 Undernutrition

Source: https://en.wikipedia.org/wiki/Malnutrition#/media/File:A_malnourished_child_in_an_MSF_treatment_tent_in_Dolo_Ado.jpg, *Creative Commons License.*

While computers have helped improve models, informatics, data management, and mapping, and molecular biologists have created new, sensitive tools to track pathogens, intervention technology has lagged behind, and prevention and control still rely on decades-old methods like nets and pesticide spraying. Despite the enormous potential of vaccinations, there have been significant problems with both the creation of vaccines and the evaluation trials that are required. For instance, vaccinations against dengue and malaria have been developed for over 30 years. There does not appear to be consensus regarding the most effective intervention strategies, even though we are starting to comprehend the effects that ecosystems and the spread of vector-borne diseases are experiencing as a result of global population growth and related human changes. Despite efforts by countries like China to halt population growth, it appears that there may eventually come a moment at which there may not be as much resource availability. Waterborne diseases and gastrointestinal infections are expected to rise with forecasted increases in heavy rainfall occurrences.

4.14 NUTRITION IS A GLOBAL CHALLENGE

Undernutrition (Figure 4.6) is still a problem, but obesity and non-communicable diseases or NCDs connected to poor nutrition now account for most of the deaths worldwide. NCDs cause 54% of lost healthy life years and account for two-thirds of all fatalities worldwide(Bennett et al., 2018). Between 2010 and 2020, there will likely be a 15% increase in annual mortality from NCDs, or 44 million fatalities. This increasing burden will mostly affect low-income and disadvantaged communities, impeding efforts to reduce poverty and promote development. Obesity, overweight, and NCDs cost society money both directly and indirectly since they raise healthcare expenditures and lower economic productivity. Over the next 20 years, the economic cost of NCDs is predicted to exceed $30 trillion, or 48% of the world GDP in 2010(Rao et al., 2019). The UNICEF conceptual framework of undernutrition emphasizes the complexity and multisectoral nature of the causes of malnutrition. Inadequate food consumption and infectious diseases are two immediate reasons that have a direct impact on the nutritional state. Due to a persistent shortage of finance and the multisectoral nature of nutritional status, eliminating malnutrition has historically been difficult. The dysfunction and fragmentation among the different actors in the nutrition community, as well as the necessity of coordination at the national and international levels, were brought to light in research papers on maternal and child nutrition.

The availability and quality of food will likely be most impacted by changes in environmental conditions, whether they come from gradual shifts in temperature and water availability or from an increase in the frequency of extreme weather events. These changes will also likely have an impact on the health of livestock in areas where they have historically been raised and the productivity of crops. If critical staple crops are produced at a lower rate due to climate change, household food security may suffer. Reduced availability of fruits and vegetables that are high in nutrients as well as higher food prices may have a detrimental impact on nutrition security, and make it harder for the poor to acquire a variety of foods. Reduced earnings and higher food prices may also have a negative effect on food access and consumption. Furthermore, the growing probability of droughts could harm food consumption by preventing access to clean drinking water. Wide-ranging impacts are anticipated from climate change, and population changes in water, air, food supply and quality, ecosystems, agriculture, and the economy may also have an indirect impact. Improving nutrition products is largely dependent on public health, particularly concerning access to clean water, sanitary conditions, and good hygiene. Populations that are impacted by climate change may have worsening nutrition as a result of a reduction in the amount or quality of water that is available. Changes in water, sanitation, and hygiene practices may also lead to decreased food utilization in addition to variations in vector-borne disease patterns. Crops that can withstand a range of environmental conditions have been created. They might not be able to withstand a mixture of interrelated environmental stressors, such as low soil fertility, heat, and drought.

Chapter 5

CLIMATE CHANGE AND HUMANS

Contents

5.1 INTRODUCTION

The anthropogenic effect must be distinguished from natural climate variability. Because of the climate system, changes made to the atmosphere now may have a lasting impact on the climate for decades or even centuries, and it is crucial. Finding out if there has been a statistically significant sign of climate change and, if so, whether any of the change can be clearly linked to human activity were two of the tasks included in the IPCC Report. One finding was that when both natural and anthropogenic effects were included in the models, there was the highest agreement between data and model simulations. IPCC concluded that human activity is primarily responsible for the majority of the warming that has been seen over the previous 50 years and that human actions will continue to alter the composition of the atmosphere throughout the twenty-first century. IPCC concluded that burning fossil fuels will almost certainly be the main cause of CO_2 emissions, which drive trends in atmospheric CO_2 concentration throughout the twenty-first century. The IPCC predicted that the global mean temperature would increase by 1.4 to 5.8 °C by the end of the twenty-first century based on climate models(Boulanger et al., 2006). Precipitation worldwide would also rise. The sensitivity of these estimates to climate change is critical. These IPCC projections are probably conservative, according to recent research that uses various methodologies and models. There is a 54% chance that the range of climate sensitivity lies outside of the IPCC's prediction, and the average global temperature may rise or fall from their estimate. There is a 40% chance that global warming will surpass IPCC projections and a 5% chance that it will fall short. Furthermore, these findings indicate that historical greenhouse gas emissions will be mostly responsible for the warming observed over the next decades, making mitigation efforts relatively ineffective.

5.2 CLIMATE CHANGE

It is anticipated that changes in the hydrologic cycle and temperature will accompany global climate change. The health impacts will also differ geographically. Higher evaporation rates will hasten the drying of soils after rain events, leading to drought conditions. In most locations where a rise in mean precipitation is predicted, including an increase in heavy rain events, larger annual variability in precipitation is expected. Other aspects of the climate system, like the frequency and strength of storms in the tropics and mid-latitudes, may also alter. The behavior of the ENSO or other types of climate variability may also be impacted by global climate change. Global mean sea level is expected to rise by 0.09–0.88 m. The main causes of this will be the thermal expansion and the mass loss from glaciers and ice caps. Because the deep ocean takes a long time to acclimatize to climate change, sea levels are expected to rise for hundreds of years after greenhouse gas concentrations stabilize. Time series studies are often utilized to analyze exposures related to short-term climatic variability. Seasonal patterns can be studied via time series analysis when assessing trends in disease rates within a single population that is defined geographically. Seasonal patterns can result from the seasonality or from other elements. Generalized estimating equations (GEE) and generalized additive models (GAM) are two commonly used methods for time series analysis(Ravindra et al., 2019). The generalized additive model uses a set of Poisson models to extract seasonal trends and long-term patterns from the data through the use of some functions. By estimating health events in connection to weather data using a regression model, the generalized estimating equation technique is comparable to GAM. For the time series, no a priori smoothing is done, though. Rather than this, the GEE model corrects autocorrelation, allowing long-term patterns in the data to be eliminated. Because the weather today is associated with the weather on the days before and after, autocorrelation needs to be taken into account when analyzing time series data of weather measurements.

Studies utilizing time series analysis were conducted in California to assess and characterize the relationships between specific meteorological factors and hospitalizations for viral pneumonia, including influenza, during normal weather and El Nino events. Analysis was done on temperature variations, precipitation, and sea surface temperature. To indicate weather variables that were not part of the analysis, like cloud cover, sea surface temperature was included. The hospitalization trends seen in some regions could not be explained by specific variations in precipitation or temperature. These findings highlight the challenges associated with attempting to model the possible impacts of climate variability on health. The objective was to ascertain if outbreaks were more closely associated with extreme precipitation incidences than with spatial clustering. Extreme precipitation occurrences were identified by using Z-score. The statistical significance was examined using a Monte Carlo model. The dangers that climate change posed to human health were not well recognized in the early 1990s. This was partly a reflection of the narrow, traditional approach to environmental health taken by epidemiologists. The environment was primarily seen as a storehouse for certain pollutants that humans had created; these pollutants might be found in food, water, soil, and the air, and each one was thought to carry a unique risk to human populations. This was made worse by a general lack of knowledge about the potential risks to long-term health that could arise from the disruption of biophysical and ecological systems. Furthermore, few physical and environmental scientists were aware that shifts in their specific research subjects like climate, biodiversity, ecosystem productivity, and so forth might have an impact on human health. The idea that human health could be impacted by global climate change was briefly discussed in the IPCC report. The IPCC does not conduct research or keep track of pertinent metrics or data about climate change. Peer-reviewed and other publicly available scientific and technical literature serve as the foundation for assessments. IPCC processes have been open to external review by other scientific bodies and government scientific advisors. National governments have authorized its work through the United Nations system. The number of reports published by the IPCC parties and the number of participating scientists both demonstrate the scope of the organization. The climate models are founded on fundamental ideas of dynamic systems and a physical understanding of the climate system. At both the global and regional scales, it is more confident to estimate climate changes resulting from a specific increase in greenhouse gas concentrations than localized variations. In the future, a number of factors will determine how quickly greenhouse concentrations really rise: levels of resource use, economic development, fertility trends, and technological advancements.

5.3 INFECTIOUS DISEASES-A SIDE EFFECT

Scientists must therefore create authentic scenarios that incorporate these aspects to serve as a guide for policymakers. Warm temperatures can result in heat stroke and cardiovascular disorders, such as heart attacks, whereas freezing temperatures can induce hypothermia and raise the risk of cardiovascular disease-related morbidity and death. Extreme rainfall, tropical cyclones, and storms can result in short-term fatalities and injuries, long-term psychological stress on impacted communities, and a medium-term increase in the risk of water-borne diseases. Gradual shifts in the climate could give human populations more time to adjust. For instance, individuals or groups may come up with innovative strategies to mitigate the effects of rising indoor temperatures. The example of vector-borne diseases provides an illustration of the complexity of interactions between environment and host. The rate at which diseases and vectors reproduce affects their ability to succeed in the environment. In a relatively short period of time, populations of mosquitoes carrying malaria can grow intensely. The Plasmodium (Figure 5.1) parasite species multiplies quickly in both human and mosquito hosts(Sato, 2021). Tsetse flies, on the other hand, reproduce slowly and take a lot longer for their populations to grow in favorable environments. As a result, unlike many diseases spread by mosquitoes, infectious diseases spread by tsetse flies respond to changes in climate more slowly. Frequency of feeding has an impact on the disease-transmission capacity of vectors. Hard ticks feed more often and for shorter durations of time than soft ticks. As a result, hard ticks are typically far more effective human disease carriers. Overall, an infectious disease with a high sensitivity to climate variability is highly susceptible to vector and pathogen reproduction, prefers humans as a source of feed, has a minimal life cycle complexity, and is highly sensitive to temperature fluctuations. Health service interventions frequently play a large role in restricting the spread of disease, even though environmental and climatic factors frequently cause changes in the rate of disease. As a result, it is frequently difficult to determine whether disease outbreaks are the result of failed interventions or changed climatic and environmental conditions. It is recognized that the climate affects infections and other health issues, it is challenging to determine the precise proportion of disease and damage that may be linked to this variable.

Figure 5.1 Plasmodium

Source: https://en.wikipedia.org/wiki/Plasmodium#/media/File:Malaria_Sporozoites_(16657642929).jpg, *Creative Commons License.*

One important means to comprehend scenarios about the potential future impacts of climate change on the incidence of infectious diseases is using mathematical models. In recent years, statistically empirical models have been employed. More integrated models are being created to consider the influence of additional factors including human behavior and the economy. Future climate and health scenarios can be inferred from historical instances of how climate variability affects health, such as the El Nino event. The conventional models of monitoring in epidemiological evaluation of diseases might not be able to keep up with the accelerated evolution of epidemics due to climate change. Often, it is challenging to determine whether an increase in malaria cases indicates the start of a widespread epidemic or just a typical seasonal variation.

The number of cases increases gradually at first before perhaps entering an exponential growth phase, which could lead to an overburden on the healthcare system. This highlights the importance of disease forecasting techniques, which can project the scale of an epidemic based on the degree of climatic anomaly. Adaptation refers to actions taken to lessen possible damage that can arise from environmental changes. Because it takes time for basic ecological systems to reach a new equilibrium, even if significant reductions in greenhouse emissions were implemented immediately, the planet would continue to heat and sea levels would continue to upsurge for hundreds of years. Many scholars are interested in learning why some systems handle stress better than others. The unusual vulnerability of certain populations to newly introduced viruses has attracted the attention of geneticists, and some have suggested that diversity (genotypes) may be a significant protective factor. The characteristics of ecosystems that are relatively resilient to stress have been studied by ecologists. Diverse plant communities/species survive in the face of extreme environmental stress, like drought, and are more successful than other species at fixing atmospheric carbon. The way that risk is distributed is also crucial.

5.4 SUSCEPTIBILITY FACTOR MODELS

Many of these concepts have been used by public health experts to examine why certain populations are more susceptible to disease than others. Famine/drought is likely to occur in populations that rely on a small number of basic foods than in those that eat a range of foods. Research on past climate-related disasters has demonstrated the critical role that political and economic systems play in influencing how extreme occurrences, such as droughts, affect health. Humans have known for ages that environmental factors can influence epidemics. Every summer, the Romans (royal family) went to their hill resorts as a means of avoiding malaria. Spicy foods in summer were less likely to cause diarrhea (South Asian). During one of the worst El Nino occurrences ever recorded in 1878, a serious outbreak of yellow fever, a virus spread by the *Aedes aegypti* mosquito, (Figure 5.2) occurred in the United States(Taylor, 2014). There was substantial human and financial damage, with an approximate death toll of 20,000 persons. It is common knowledge in developed nations today that influenza epidemics frequently return in the middle of winter. Transmission of infectious diseases needs to be understood in the context of ecology. Infectious agents parasitize higher organisms to gain the nutrition and energy they need. A small percentage of infections that negatively impact the host are classified as infectious diseases. The patterns of infectious disease in humans have changed due to several transitions in human ecology and inter-population interactions, which have occurred over the long processes of human evolution, population spread, and subsequent contact and conflict. The related vector organisms like mosquitoes, ticks, and sandflies, and infectious agents like protozoa, bacteria, and viruses, are both small and lack thermostatic mechanisms. Therefore, each species has a specific set of climatic conditions within which it may survive and proliferate. It is also important that the incubation period within its vector is usually sensitive to temperature variations, typically exhibiting a connection. The amount of precipitation, sea level rise, wind, and length of sunlight are additional climate sensitivity factors for the agent, vector, and host.

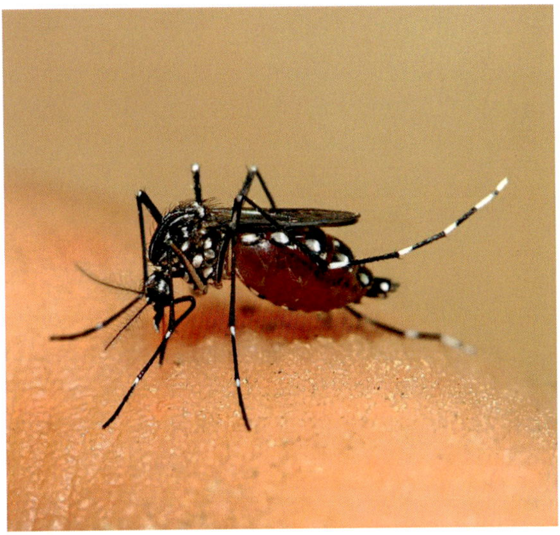

Figure 5.2 *Aedes aegypti*

Source: https://en.wikipedia.org/wiki/Aedes#/media/File:Aedes_aegypti_feeding.jpg, *Creative Commons License.*

Among the many significant variables affecting the prevalence of infectious diseases is the climate. In addition, sociodemographic factors like drug resistance, nutrition, and human migration and mobility, as well as environmental factors like deforestation, agricultural development, water projects, and urbanization, should be taken into account. It is exceedingly improbable that climatic changes have an isolated effect on disease in this era of global expansion and shifting land-use; rather, the effect is probably influenced by how well humans adapt to or counteract the trends of other factors that modify disease. We need to look at how much these non-climatic factors may exacerbate the effects of climatic circumstances on diseases, while acknowledging their individual/separate importance. Specialists can categorize infectious diseases using multiple approaches. The clinical manifestation (disease) is the most important factor for professionals who treat people. On the other hand, microbiologists typically use the traits of microorganisms to categorize infectious diseases. Since epidemiologists are primarily concerned with controlling the spread of disease and preventing future outbreaks, the pathogen's natural reservoir and mode of transmission are important properties. The influence of climate change on diseases is mostly dictated by the distinct cycle of transmission of pathogens. When a disease involves both a human and a vector, its transmission cycle is more vulnerable to outside environmental factors than when it solely involves the pathogen and the human. Temperature, precipitation, and humidity are important environmental influences. Pathogens like viral, bacterial, etc., vectors like mosquitoes, snails, etc., non-biological physical vectors like water, soil, etc., non-human reservoirs like mice, deer, etc., and human hosts are a few potential components of transmission. Infectious diseases are generally categorized by epidemiologists as either anthroponoses or zoonoses, depending on the pathogen's natural reservoir; additionally, direct or indirect infections are identified based on the mode of transmission. Diseases known as directly transmitted anthroponoses are ones in which the pathogen is typically spread directly between two human hosts by touch or contact with droplets. The pathogen and the human host are the two components that make up the transmission cycle of these diseases. Since the agent spends little to no time outside the human host, climate conditions are generally least likely to have an impact on these diseases. These diseases are vulnerable to alterations in human behavior brought about by factors like overcrowding and poor sanitation, which may arise from modified land use brought on by climate change. Measles, tuberculosis, and sexually transmitted infections like herpes, syphilis, and HIV are a few examples of directly transmitted anthroponoses(John et al., 2011).

5.5 TRANSMISSION AND TRIGGERS

Similar to directly transmitted anthroponoses, directly transmitted zoonoses are spread through physical contact or exposure between reservoirs. Nonetheless, it is thought that an unintentional human contact is what caused the infection of humans, as these agents are naturally dispersed across animal reservoirs. These diseases are more vulnerable to the effects of climate variability because the persistence of these pathogens in nature is primarily dependent on the interaction between the animal reservoir and the external environment, which can affect the rate of transmission, host immunity, rate of reproduction, and rate of death. The hantavirus is a naturally occurring, directly transmitted zoonosis that can spread to humans during periods of elevated local reservoir abundance. It is sustained in rodent reservoirs. Another directly transmitted zoonosis that affects small mammals spontaneously is rabies. It is extremely harmful to its vertebrate host and has very little chance of spreading widely. Many manmade diseases that affect people today, like HIV and tuberculosis, were first discovered in animals. A family of diseases known as indirectly transmitted anthroponoses is characterized by the transmission of pathogens between two human hosts by either a biological vector, such as ticks, or a physical carrier, such as soil. A full cycle of transmission for these diseases needs three elements: the infection, the physical or biological vector, and the human host. For most vectors to survive and reproduce, they need to consume blood from their vertebrate host. Malaria and dengue fever are examples of indirectly transmitted anthroponoses, in which mosquito vectors transfer the corresponding malaria parasite and dengue virus among human hosts. Climate impacts can affect water-borne anthroponoses that are spread indirectly since the pathogens spend a portion of their life cycles in the external environment. Flooding may contaminate water sources, or ambient air temperatures may have an impact on the pathogen's rate of reproduction. *Vibrio cholerae* is spread through waterborne vectors and is an indirect water-borne anthroponose. The bacteria live in marine habitats where they attach themselves to zooplankton. Furthermore, the quantity of phytoplankton in their food source determines the survival of these small crustaceans. Warm ocean temperatures typically cause phytoplankton populations to flourish(Sarker et al., 2020). These ecological linkages explain why rising ocean surface temperatures trigger cholera outbreaks. Similar to indirectly transmitted anthroponoses, indirectly transmitted zoonoses are spread naturally amongst non-human animals, with people becoming infected by unintentional contact with an infected host or vector. Four elements make up the transmission cycle for this class of disease: the pathogen, the biological vector or physical vector, the animal reservoir, and the human host. Due to the multiple elements in the transmission cycle and how each interacts with the surrounding environment, these diseases are extremely vulnerable to a confluence of ecological and climatic conditions.

For many diseases that cannot be categorized solely by natural reservoir or mode of transmission, complex cycles of disease transmission also exist. One such disease is Rift Valley fever, which is caused by a zoonotic virus that is carried by the Aedes mosquito species and spreads across vertebrate hosts. Culex mosquitoes can feed on infected hosts, especially during floods. Because it feeds on humans as well, this vector is known as a bridge species because it spreads the virus outside of the regular zoonotic cycle. This disease pattern could be the consequence of indirect social or behavioral adaptations to the cold, including crowding indoors, which enhances the risk of transmission. Another explanation is that bacterial sensitivity to environmental variables like humidity may be the direct cause. Apart from influenza, many other infectious diseases display seasonal patterns that are cyclical and could be attributed to the climate. Enteric infections exhibit prominent seasonal variations in a variety of locations. Scotland experiences brief springtime surges in campylobacter infections. Cholera outbreaks occur in Bangladesh during the monsoon season. Infections with cyclospora rise in the summer and decrease in the winter in Peru. Similar to this, there are prominent seasonal variations in the transmission of certain vector-borne diseases, such as dengue fever and malaria, with the highest rates occurring during the months with high humidity and rainfall. In sub-Saharan Africa, epidemics of other diseases, such as Meningococcal meningitis, typically spread during the hot, dry season and disappear shortly after the start of the rainy season. The seasonal differences in the incidence of infectious diseases suggest a relationship with meteorological conditions. However, non-climatic factors need to be taken into account to establish a causal relationship with climate. Furthermore, data from multiple seasons must be used, together with appropriate statistics to take seasonal changes into account, to evaluate the long-term effects of climate change on disease trends.

5.6 VECTORS AND PATHOGENS

Although disease-causing pathogens are frequently killed by extreme temperatures, small temperature changes can have a variety of impacts. A slight temperature rise may be fatal to the pathogen in a vector habitat where the mean temperature is close to the threshold of the pathogen's physiological tolerance. On the other hand, if a vector resides in an area with a low mean temperature, even a slight rise in temperature may cause the disease to grow, incubate, and replicate more quickly. Temperature can affect the dynamics of vector populations, change the rate at which they come into contact with humans, and modify the growth of disease-carrying vectors by changing their bite/contact rates. The duration of the transmission season may change due to a change in the temperature.

Vectors that spread disease can adjust their geographic distribution in response to temperature variations. Mosquito populations residing at higher latitudes were shown to exhibit a higher degree of microevolutionary response; this suggests that these populations are more likely to undergo genetic evolution due to increased selection pressure. Even though some studies were restricted to a single species of mosquito, it raises the possibility that other mosquito species, possibly those that transmit diseases, may go through a similar microevolution that would enable adaptation to different seasonal patterns brought on by global climate change. Especially for insect vectors, humidity can have a significant impact on the spread of diseases carried by vectors. In dry conditions, the survival rate of ticks and mosquitoes falls because they can quickly die. It has been discovered that one of the most important factors in climate/disease models, such as dengue fever and Lyme disease, is water deficit, which is comparable to relative humidity. The anticipated elevation in sea level linked to climate change is probably going to reduce or eradicate saltmarsh mosquito breeding. Extinction of the bird and mammalian hosts that fill this ecological niche could be a hazard, which would help eradicate the viruses that are exclusive to this environment. Inland saltwater intrusion may convert parts of freshwater habitat into saline regions, supporting vector and host species that have been removed from their previous saltmarsh habitats(Russell, 1998). Contact with contaminated drinking water or food can expose humans to water-borne diseases. Exposure can result from weather-related phenomena or human operations, such as improper disposal of sewage waste. Temperature has an impact on the growth and survival of pathogens, and rainfall patterns can affect how those microorganisms travel and spread. Warming temperatures have the potential to extend the seasonality of water-borne diseases or change their geographic distribution. Warm temperatures in the marine environment are conducive to red tides, which are toxic algal blooms that can raise the risk of shellfish poisoning. By augmenting the food supply in their reservoir, rising sea surface temperatures can have an indirect effect on the survival of enteric pathogens like *Vibrio cholerae*.

Rainfall can contaminate watersheds by introducing other wastes including human and animal feces into the groundwater. Giardia, E. coli, and Cryptosporidium have all been linked to water contamination after severe/heavy rains. Because of more effective microbial transport, this kind of occurrence might occur frequently in high soil saturation conditions. In poorer countries, lack of water has been linked to a rise in outbreaks of diarrhea, which is probably caused by poor hygiene. There is evidence linking certain arbovirus epidemics, like the Saint Louis encephalitis virus (SLEV), to environmental variables. Every SLEV-based event took place during a rainy season that was immediately succeeded by a drought. A perfect environment is created for the quick spread and multiplication of SLEV by the forced interaction between mosquito vectors and bird hosts. When the drought breaks and water supplies are replenished, the SLEV-transmitting vectors and hosts spread over a larger geographic region. A comparable impact of climate change has been proposed for the West Nile virus (WNV). A mild winter followed by hot, dry summers is thought to be the climate factor that causes the virus to multiply. WNV is a vector-borne zoonotic disease, much like SLEV, that is typically spread between birds by the *Culex pipiens* mosquito(Ciota & Kramer, 2013). The vector prefers contaminated stagnant water for breeding. Drought causes this water to concentrate organic matter even more, and mosquito predators like frogs become less common. Birds may move around small watering holes, increasing mosquito attacks. Such weather patterns were observed in some areas. These could be explanations for the epidemic, along with urban and suburban settings that are suited for bird species. There is scientific information/data indicating that malaria varies seasonally in highly endemic location. The vector-borne disease susceptible to long-term climatic change is most likely malaria.

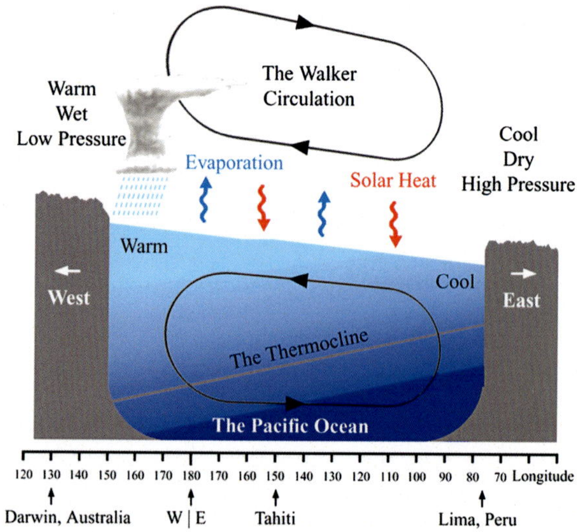

Figure 5.3 ENSO cycle

Source: https://en.wikipedia.org/wiki/El_Ni%C3%B1o%E2%80%93Southern_Oscillation#/media/File:LaNina.png, *Creative Commons License.*

5.7 SPREAD OF DISEASES

Many countries have looked into the connection between malaria and interannual climate variation linked to the ENSO cycle (Figure 5.3). For instance, in an El Nino year, Venezuela receives less rainfall. The year after an El Nino year saw a rise in malaria rates of more than thirty percent. The probable causes of this include the combination of heavy rainfall in the year following the El Nino and mosquito-favoring conditions, together with temporarily lowered immunity levels in the local population after the previous year of low incidence. The majority of correlations that have been shown between climate change and water-borne diseases rely on evidence of seasonal variations. Nonetheless, a number of studies offer quantitative proof of the connections between precipitation and ambient air temperature and water-borne diseases. Some researchers studied the relationship between high precipitation and outbreaks of water-borne diseases in the US between 1948 and 1994. Their results demonstrate a statistically significant correlation between outbreaks of water-borne diseases and high precipitation, with over 50% of outbreaks being preceded by exceptionally rainy months falling within the upper tenth percentile of a 50-year record of monthly rainfall. There is ample evidence of numerous biological and physical indications of the long-term impacts of climate change. A variety of biological and physical processes influence human health. If disease does exist, there are still many barriers to identifying reliable trends in the incidence of human disease, including variability in detection and/or reporting. Many studies have found links between the occurrence of infectious diseases and long-term trends in climate change, despite the challenges that hinder quantitative studies of these relationships. Researchers observed a correlation between the occurrence of cholera in Bangladesh and increasingly greater El Nino occurrences. The researchers conducted a time series study of historical cholera data using statistical techniques to observe the impact of interannual variability that may be related to climate change. The El Nino Southern Oscillation (ENSO) has changed during the past 20 years compared to earlier decades. The scientists discovered that while the relationship between cholera incidence and ENSO was weak and uncorrelated in the first half of the century (1893–1940), it was significant and persistent in the latter half of the century (1980–2001)(Shackleton et al., 2023). Therefore, past climate change may have already had an impact on regional cholera trends through enhanced ENSO events. Relationships between vector-borne diseases and climate may be associated with other variables like socioeconomic, demographic, and environmental factors. Models are useful tools for analyzing the intricate processes by which infectious diseases spread. By using global or regional climate models, they enable researchers to connect future spatially gridded estimates of climate change to a model of the association between those climatic factors and the occurrence of the disease of interest. Models are probably going to play a great role in both guiding future research and forecasting disease risk. Although the intricate modeling needed to research infectious diseases is difficult, it has been done successfully on many occasions.

5.8 METEOROLOGICAL FACTORS

A range of meteorological factors have been integrated into empirical models to characterize the bioclimate or climatic restrictions, for many vector-borne diseases. The evaluation is based on an ecoclimatic index that is mostly influenced by the vector's needs for moisture and temperature. According to CLIMEX calculations carried out in Australia, in a typical climate change scenario, the native malaria vector might extend its range by 330 km south(Molina-Guzmán et al., 2022). These approaches, however, are unable to account for every element influencing species distributions. The amount to which a species colonizes a suitable environment depends on several factors, including competition and interaction between species as well as local topographical barriers. Additional dynamic population models like DYMEX, may also be included in assessments. Some models were used to estimate variations in malaria risk based on humidity and the minimum/maximum temperatures required for parasite development, together with general circulation models (GCMs) of the climate. These models took into account regions where malaria had been eradicated and very accurately reflected the spread of the disease in the 1990s. The fact that every simulation run indicated a rise in seasonal malaria transmission was a significant finding. Using a two-step statistical method, they were able to forecast future falciparum distributions by using current falciparum malaria distributions to identify significant climatic elements and projecting future falciparum distributions using this information. Future falciparum habitat distributions were predicted by the model to resemble those of the current. One of their findings is that other variables, like precipitation, would probably balance the organism's biochemical response to temperature extremes. Some researchers have employed an alternative method that combines experimental and observational data to investigate the impact of temperature on the survival of the *Culex tarsalis* mosquito species, which is the main vector of Western Equine Encephalitis (WEE) and Saint Louis encephalitis (SLE)(Tsai & Mitchell, 1988). Vector populations peak in the spring in the southern region and in the middle of summer in the northern regions.

5.9 PREDICTIVE FACTORS

Climate affects the rate of biological processes of pathogens and vectors directly, as well as the habitats in which they live, which in turn affects the spread of infectious diseases. A vector, host, or pathogen's environmental constraint can be determined using satellite data. The important elements influencing the existing distribution of diseases and the impact of anticipated climatic changes on this distribution can be identified by analyzing them with meteorological data. Many investigations on malaria have made use of remote sensing. Temperature and moisture levels affect mosquito dispersal and the length of time malaria takes to incubate. This makes it possible to forecast mosquito distribution and transmission patterns using climate data from satellite imagery. These data in conjunction with internal variables like population nutrition and immunity, can aid in the development of localized models suitable for current climatic circumstances. These in turn can serve as warning systems in the future for anticipated climatic scenarios. A key example of this kind of modeling is the MARA project, which stratified malaria risk in the African highlands using satellite data. Some researchers investigated the effect of mosquito net use on child malaria prevalence rates in the Gambia using satellite-based geographical analysis. To forecast the duration and severity of malaria transmission, changes in vegetative growth and senescence are measured using satellite data, which serves as an ecological variable for variations in humidity and rainfall. Using multiple regression analysis, the researchers discovered a substantial correlation between the prevalence of malaria and both the NDVI and the use of nets, whether treated or not, after taking into account the spatial correlation of the villages included. After considering behavioral factors, studies indicated that the NDVI data can be a helpful predictor of the prevalence of malaria in children. 90% of the peak concentrations of larvae were predicted by the satellite data based on the distance between livestock pastures and rice fields. The ecology of the main malaria vector, *Anopheles albimanus*, (Figure 5.4) in the Mexican state of Chiapas was examined. Specifically, the relationship between mosquito larval populations and aquatic plant species was examined using digitally processed satellite data. The findings indicated that the two most significant landscape elements linked to vector abundance were transitional marshes and unmanaged pastures.

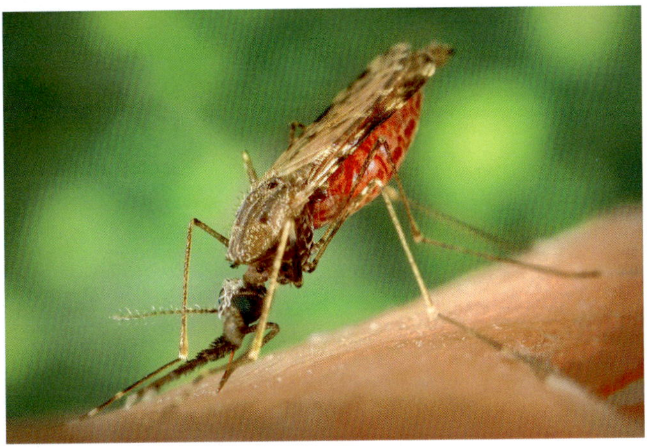

Figure 5.4 *Anopheles albimanus*

Source: https://en.wikipedia.org/wiki/Anopheles_albimanus#/media/File:Anopheles_albimanus_mosquito.jpg, *Creative Commons License.*

Researchers examined the relationship between vegetation and specific Belizean settlements with high or low mosquito concentrations using a combination of digital maps, remotely sensed data, and field data. The value of remotely sensed data for malaria surveillance is demonstrated by many investigations. Using remote sensing data, several vector-borne diseases have been investigated. Three significant disease vectors like mosquitoes, tsetse flies, and ticks were mapped in southern Africa under different climate change scenarios. The findings show that there could be major changes in the regions that are appropriate for each species of vector, with *Anopheles gambiae* mosquitoes showing a net increase. Scientists mapped the geographic range of *Phlebotomus orientalis*, the sandfly vector that transmits visceral leishmaniasis in the Sudan, using satellite data. GIS was used to extract data on environmental variables related to the abundance of tick populations on white-tailed deer, the zoonotic host, the tick vector that spreads Lyme disease in the eastern United States. Tick abundance was shown to be negatively connected with privately held land, wetlands, urban land-use patterns, saturated soils, and limited drainage, and positively correlated with well-drained sandy soils. Similar findings in the distribution and abundance of *Ixodes scapularis* in Wisconsin, northern Illinois, and parts of Michigan were demonstrated. ENSO-based disease forecasts can benefit from new information procured from observations of the unexpected climate variability that accompanies the El Nino phenomenon.

Human instances of Hantavirus pulmonary syndrome (HPS) in the U.S. were preceded by excessive rainfall associated with ENSO, which was followed by an increase in the rodent population(Douglas et al., 2022). By using logistic regression analysis and Landsat Thematic Mapper satellite imagery HPS risk was assessed. The next year, there was a clear correlation between environmental factors and HPS risk, as evidenced by satellite data. Satellite imagery from 1995, a year without an ENSO, was used for analysis, and the results showed a significant decline in medium to high-risk locations. This ENSO-based forecasting model is part of a regional early warning system for hantaviruses used by the health services in the United States. In East Africa, ENSO-driven rainfall and Rift Valley fever have a high correlation. Between 1950 and 1998, episodes of heavy rainfall were followed by outbreaks of the Rift Valley fever virus in East Africa. It was shown that Rift Valley fever outbreaks might be anticipated up to several months ahead of outbreaks in East Africa by combining NDVI data with sea surface temperature anomalies from the Pacific and Indian Oceans. Real-time tracking using satellite imagery could identify the actual impacted regions, and this could prove to be a valuable disease management tool. The World Health Organization has expressed the necessity for an operational early warning system that would allow the mobilization of control activities with enough lead time. Sea surface temperature anomalies and monthly dengue cases in many cities are predictive variables. Persistent international dedication has played a major role in the successful eradication and management of infectious diseases like polio and smallpox. In other cases, inconsistent efforts to control the disease have had negative effects on human infection. The Pan-American Health Organization (PAHO) launched an initiative in 1947 intending to eliminate *Aedes aegypti* (yellow fever)(Lopes & Reis-Castro, 2019). An infectious agent becomes resistant to drugs by genetically changing its genome to evade the harmful effects of the drugs. Since the 1970s, resistance to significant anti-malarial drugs has increased in various regions of the world. Resistance to *P. falciparum* has expanded geographically throughout Africa as a result of the use of chloroquine as a treatment for malaria infections. Similarly, the widespread use of antibiotics in humans and livestock has caused bacterial pathogens to become resistant, which is a major issue in poor countries without access to medical supervision for the proper dispensing of medicine/drugs. The mortality of infectious diseases is significantly influenced by malnutrition. Micronutrients are crucial for the body's capacity to establish a successful defense against disease. In children in poor countries, malnutrition is a significant risk factor for diarrhea. Nevertheless, unexpected consequences could include detrimental effects on human health in addition to negative environmental effects.

5.10 OTHER KEY CHANGES

The conversion of forests for grasslands or crop production, the construction of roads or dams, the logging industry, and the expansion of urban areas are all factors endangering the condition of the forests. The diseases that commonly harm animals are those that naturally occur in wild situations and spread among them, particularly those that have invertebrate vectors and vertebrate reservoirs. Typically, these alterations stem from elements impacting the populations of diseases, vectors, and animal reservoirs, as well as elements linked to human exposure. Removing vegetation can drastically change the local weather patterns by affecting plant transpiration. Climate change-related floods are predicted to worsen trends in water contamination brought on by agricultural growth. Many regions have seen an increase in crop irrigation due to agricultural growth, which has also reduced the amount of water available for other purposes and increased the number of places where disease vectors can survive. Following the completion of the Aswan High Dam, the Southern Nile Delta had an increase in soil moisture due to irrigation development. This led to a rapid growth in the *Culex pipiens* mosquito population, which in turn produced an increase in the mosquito-borne disease *Bancroftian filariasis*. Other vector-borne parasite infections that are brought on by shifting patterns of land use and water management include Onchocerciasis and Trypanosomiasis. Droughts will force additional irrigation to counteract water stress, which may result in an increase in the number of vector-borne disease breeding grounds.

A number of health issues are linked to urbanization, such as respiratory disorders, diarrheal and vector-borne diseases like dengue and malaria. These conditions can be due to pollution brought on by poor infrastructure and overpopulation. Water-related diseases can be significantly decreased by implementing water sanitation measures. Apart from alterations in land use, there exist several indirect connections between infectious diseases and environmental conditions. These connections are mediated by alterations in ecosystems that arise from human activities. Because they depend on the ecology of non-human animals, anthroponoses and zoonoses spread via vectors will be vulnerable to the consequences of ecological shifts. Since exposure to zoonotic infections is thought to have developed into 75% of new infectious diseases in humans, alterations in the ecological factors influencing diseases in wildlife could have a direct effect on human health. Changes in host sensitivity to diseases, seasonal patterns of pathogen survival, and the timing of pathogen development and life cycles are all effects of climate change on the connections between pathogens and hosts. However, there is an indirect relationship between ecosystem processes and infectious human diseases. The challenges of projecting how ecosystems would react to climate change affect predictions of the responses of infectious diseases to it. Temperature and precipitation patterns influence the geographic distributions and species abundances that make up ecological communities. Because metabolic rates and temperature are correlated, average temperatures will also affect many species' growth rates. Precipitation patterns also affect species distribution and abundance by defining the range of plants or habitats or the locations of vectors (breeding). Organisms responsible for human infectious diseases will be impacted by changes in the aquatic ecosystem because mosquitoes that spread dengue fever and malaria breed in aquatic environments.

5.11 RESPONSES AND TRENDS

Predicting the response of organisms is one of the key issues in ecology. In an early effort to identify potential areas for the development of favorable conditions for a species, data on species tolerance to various climates and their present geographic distributions have been combined with estimates from models. The degree to which the existing distributions of tree species were associated with the current climate differed among the species. The expected effects of future climate change varied throughout species as well. Comparable trends emerge when anticipating alterations in animal communities. The capacity of different species to spread to new regions and face competition from established species will also differ. While some species can only spread slowly, others can quickly change. The capacity of different species to establish themselves and grow their populations after arriving in a new region varies. Many species have the essential genetic variety, but environmental change may happen too quickly for them to adapt(Jump & Peñuelas, 2005). A deeper comprehension of evolutionary function in disease ecology is an important topic of study. Ecosystems will change, even if there are a lot of unknowns regarding the precise type of ecological changes that are anticipated as a result of climate change. In fact, there have been important alterations to the phenology of species. Common changes in the abundance and phenology of species have been found in analyses of ecological data. These modifications have been linked to the relatively mild climate changes of the previous years. It seems that some species have undergone adaptive evolution in response to the recent climate shift. As predicted temperature changes persist over the next decades, changes in ecological networks will become more apparent. There is strong evidence that many species will react to climate change at different rates and with diverse degrees of sensitivity. Important ecological networks are expected to be disrupted by these irregular responses.

In comparison to other health risk factors, climate change is unique in that a substantial amount of work has gone into creating and assessing models to predict future climate in response to predicted trajectories of atmospheric gas variations. Information has been grouped regarding the connections between many important health effects and variations in the climate, whether they occur geographically or over short periods. It offers a starting point for estimating the magnitude of climate change effects on various scales. It is predicted that climate change will have an effect on how deaths resulting from exposure to high or low temperatures are distributed i.e., reduced mortality in winter, but increased summer mortality. According to estimates, in 2030, the risk of diarrhea could increase by up to 10% compared to the case of low climate change(Ebi, 2008). These uncertainties are primarily related to the limited studies that have described the exposure-response relationship. Regional estimates of the effects of malnutrition vary significantly. The relative hazards of uncontrolled emissions compared to low climate change by 2030 range from a slight decline in the western Pacific region to a major increase in Asia (southeast). With discontinuous stabilization of climate change, there is no discernible trend of declining relative risks. These estimates correspond to a substantial disease burden, although appearing unstable due to their sensitivity to regional variation in precipitation. Although there are proportional variations in the number of fatalities from coastal floods, the numbers of casualties are quite low. It is anticipated that the effects of inland floods will worsen resulting in a higher burden of acute disease. There is uncertainty in the estimates of the efficacy of adaptation strategies as well as the quantitative correlations between variations in precipitation, flooding, and the resulting health effects. Precipitation changes are unpredictable, which is most likely why certain places appear to show a tendency toward decreased incidence with rising GHG.

Chapter 6

HUMAN ADAPTATION

Contents

6.1 INTRODUCTION

Variations in climate occur at every temporal and spatial scale, ranging from the regional, continental to the global. For instance, although the Southern Oscillation in the atmosphere that is linked to the oceanic El Nino phenomenon is visible in the tropical Pacific (Peru), its effects are almost global. In general, natural climate variability refers to variations that are caused by random fluctuations in physical or chemical factors. Decadal to multi-decadal climate variability can be established based on records of surface air temperature, sea surface temperature (SST), (Figure 6.1)and pressure(Mohino et al., 2011).

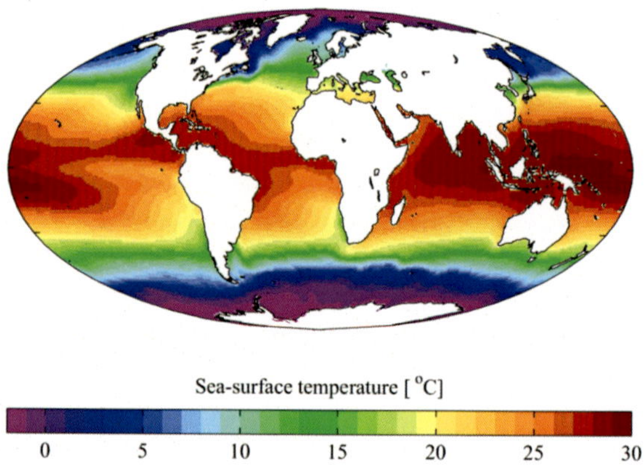

Figure 6.1 Sea surface temperature

Source: https://simple.wikipedia.org/wiki/Sea_surface_temperature#/media/File:Sea_surface_temperature_2001.png, *Creative Commons License.*

Indexes measuring variations in a certain atmospheric or oceanic parameter, such as sea surface temperature or pressure, within an area are commonly used to characterize large-scale oceanic–atmospheric oscillations that impact weather (Global or regional). El Nino–Southern Oscillation (ENSO) has an impact on the global climate. The weakening/strengthening of the winds over the tropical Pacific is what defines the Southern Oscillation, an interannual oscillation in tropical sea pressure between the western and eastern Pacific. El Nino and the Southern Oscillation are known to be closely related. Long-term temperature trends indicate rising up to roughly 50 Ma and then long-term cooling that continues into the Holocene. The Paleocene–Eocene Thermal Maximum, which occurs at around 56 Ma, and the Middle Eocene Climatic Optimum, which occurs at approximately 42 Ma, are prominent examples of intermittent global warming spikes. These events correlate with temporary increases in atmospheric CO_2. Large ice sheets grew across the Pliocene–Pleistocene transition, causing severe glacial advances in the Pliocene Arctic and intensification of the Northern

Hemisphere glaciation. This irreversibly changed the northern areas/regions. Over the last 800,000 years, atmospheric CO_2, CH_4, and N_2O have changed simultaneously with global temperature(Schilt et al., 2010). This interpretation is based on accurate data retrieved from ice cores. The solar energy received at the TOA or top of the atmosphere and its distribution across latitudes are modulated/altered by the slowly changing characteristics of the Earth orbit.

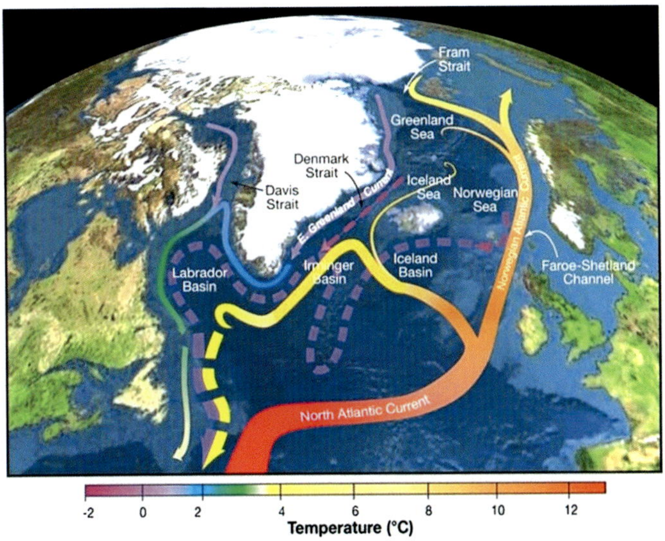

Figure 6.2 Atlantic meridional overturning circulation

Source: https://en.wikipedia.org/wiki/File:OCP07_Fig-6.jpg, *Creative Commons License.*

6.2 THE ATLANTIC MERIDIONAL OVERTURNING CIRCULATION

The Atlantic meridional overturning circulation or AMOC (Figure 6.2) currently serves as an important example of the significance of oceanic circulation in regulating our climate. This current system plays an important role in monitoring the cycling of carbon and other nutrients as well as the distribution of heat on Earth. It is distinguished by the deep-water formation in the North Atlantic, cooler water masses in the deep Atlantic Ocean, and the warm northward-flowing Gulf Stream in the upper layers of the Atlantic Ocean. The near-surface waters that flow northward effectively contribute to the transmission of heat from the Tropics to the mid- and high latitudes, hence influencing the temperature in the surrounding region of the North Atlantic. Circulation and convection in the North Atlantic have fluctuated during the Pleistocene, which corresponds to warm and cold climates. Both within the glacial phases of these cycles and at the glacial-interglacial cycles, substantial changes took place. Even though the Holocene is thought to have had a stable climate, considerable climate variations that are high/maximum have been documented in the natural archives (glaciers/rocks). The most notable of these changes include the Little Ice Age, the Medieval Climate Anomaly, the Holocene Thermal Maximum and rapid deglaciation, declining insolation since five millennia ago, and the variability of the monsoon linked to aridification. The capacity of climate models to replicate the observable paleoclimatic records can be useful in elucidating the mechanisms underlying the Holocene climate variability.

6.3 CLIMATE EVENTS

There are two epochs within the Quaternary, the Pleistocene and the Holocene, which span 2.6 million years ago to the present period. The Holocene, the beginning of the current interglacial, was determined to have begun 11,700 years ago(Borzenkova et al., 2015). It is interpreted based on isotopic evidence from the Greenland NGRIP ice core. Prior to the 1950s, the Blytt-Sernander scheme used peat stratigraphy to divide the Holocene into climatic intervals/periods. A distinction is made between the start of the Holocene, which is defined as 11.7 ka, and the beginning of interglacial conditions, which is determined by looking at pollen records from Northern Europe and occurring at 10 ka. It was suggested that the Mid to Late Holocene Boundary be located about 4.2 ka. The scientific community can gain a better knowledge of climate variability by utilizing Earth system models (ESM). Simulations involve forcings and boundary conditions, which can change. Volcanic forcing, variations in solar radiation, and the orbital forcing have been the core factors across the Holocene. The implication of anthropogenic land use and greenhouse gas emissions increases as the industrial age draws close. One of the main natural sources of climate change is volcanic eruptions.

Their injections of sulfate aerosol into the stratosphere provide a barrier from incoming solar radiation. This will cause short-term cooling at the regional level (can extend to the global level too). The history of solar events over an extended time span can be revealed by long-term records of isotopes such as $_{14}C$ and $_{10}Be$, even when direct solar observations are limited to the previous decades. In general circulation models (GCMs), we can observe the fundamental processes like convection and cloud formation, which have long been recognized as essential to the troposphere energy dynamics. GCMs have progressively evolved into Earth system models. The monsoon climate affects around half of the population globally and occupies about a quarter of the continents. Interannual and interdecadal variability in precipitation is a distinctive feature of the Sahel region of Africa up to eastern China. This resulted in sporadic floods and droughts. Many people died in rural areas in the Sahel during a drought. Many deaths were attributed to starvation, malnutrition, and infections. Devastating droughts were documented in China due to monsoon failure. A lot of research was done to anticipate/forecast precipitation and reconstruct monsoon variability during the Holocene period. The retreat of the global monsoon towards the equatorial zone in Africa and Asia, and the consistent aridification of the regions north of the affected area, are thought to be the most important climate changes that have occurred during the Holocene period. It left several severe implications for prehistoric civilizations.

6.4 MODELS

Using a low-resolution general circulation model, researchers carried out experiments to investigate the impact of Earth's orbital parameters during the Early Holocene period. The increase in solar radiation in this simulation experiment causes the land surface to warm in relation to the ocean. It was interpreted that there was intense summer monsoon circulation over the African-Eurasian areas. This is closely related to paleoclimatic evidence that shows conditions existed in parts of Africa, Arabia, and India between 10 and 5 ka ago. Paleoclimate research conducted in the last ten years has yielded unprecedented temporal and spatial coverage of previous monsoon variability. The global monsoon concept considers the cohesive response of all monsoon systems, irrespective of regional disparities.

The subsystems of North America, South America, North Africa, Southern Africa, South Asia Monsoon, East Asia Summer Monsoon, and Australia–Indonesia build up the global monsoon, which is defined by the seasonality in rainfall mostly. Given its potential consequences for the evolution of African hominids and other species, some researchers delved deeper into the hydrological history of North Africa throughout the Pleistocene periods. The East Asian Summer Monsoon or EASM and Indian Summer Monsoon (ISM) are the two subsystems that create the Asian summer monsoon (ASM)(Wang et al., 2010). These subsystems are responsible for most of the rainfall in China. India experienced substantial monsoon fluctuation during the Holocene (based on paleoclimatic data). The central Indian flora is dated by carbon isotopes of sedimentary leaf waxes from a sediment core recovered from the mouth of the Godavari River (Bay of Bengal). These isotopes showed an increase in desert-adapted vegetation from 1,700 years ago. Six periods of an enhanced Indian Summer Monsoon (ISM) were observed in the central Himalayas. It was shown that the drift-ice record of the northern Atlantic and the phases of weakened ISM largely correlate. This suggested a coupling between the multi-centennial-scale monsoon variability in the central Himalaya and the short-term solar radiation-induced glacial boundary condition in the northern Atlantic.

6.5 MONSOONS

The interplay between the mid-latitude westerlies and the East Asian summer monsoon was identified. It was demarcated from the transitional climatic zone (TCZ) (between wet and arid regions in East Asia). This zone is marked by severe climate and biome variations. It runs from the Tibetan Plateau to the western region of northeastern China, with humid climate conditions to the SE and NE. Some researchers aimed to investigate the underlying mechanisms of drought variability at long and short time-scales in the northern frontier of the Asian summer monsoon region (NASM). This is a critical area of study due to the impending scarcity of water. The region, which is sensitive to climate variability in the transition zone from the monsoon-dominated area to the westerly dominated NASM, has had multiple drought events (past 500 years). Similar monsoon patterns have been observed in the Cis-Baikal region as a result of a change in climate and the departure of the Kitoi culture during drought between 6-7 ka during the southward shift of the EASM. The pattern that people react to changes in the climate and environment varies at all levels. They might relocate, adjust, develop appropriate technology, or see some changes in their economic and sociocultural systems. Humans can adapt to changed environments. We can observe cultural change as a common response. The social and economic conditions were governed by the ecology. These conditions will alter to changes in the neighboring conditions to maximize survival rates. This could require adapting while migrating or adapting in place. Understanding how humans have historically adapted to rough environmental conditions is crucial to understanding our survival instincts. Hunter-gatherers settled in high-altitude Andean habitats during the late Pleistocene, despite the cold and low oxygen levels. There is evidence of human presence in both southern and northern Scandinavia in the archaeological record (11,700 years ago)(Wygal & Heidenreich, 2014). Scandinavia was dominated by the ice sheet at that time. Researchers combined genetics, climate modeling, and archaeology to investigate the postglacial migrations into Scandinavia. This revealed adaptation strategies in high-latitude environments (Mesolithic period). Researchers examined many radiocarbon dating records from archaeological sites in Britain and Ireland (Holocene). They found that there were multiple periods of population decline. These periods also seem to be linked to known solar grand minima and episodes of rapid Holocene climate shifts to unstable weather scenarios in the North Atlantic.

6.6 SOCIAL FACTORS

Some researchers developed a conceptual model that incorporated climatic and social factors to describe the long-term dynamics of migration (farmers) in China over the last two millennia. The model was based on data of migration and social emergency events, such as wars, famines, and epidemics. The statistical research explains why farmers are reluctant to migrate. Farmers would have remained where they are if climate change emerged. However, whenever a social disaster threatened to endanger their lives, they fled their houses. Europeans responded to climate change between 1500 and 1800 AD(Raible et al., 2006). This response demonstrated how cooling between 1560 and 1660 AD resulted in disasters like agroecological, socioeconomic, and demographic disasters that eventually led to the Crisis of the 17th Century. Scientists discovered causal associations between human disasters and climate change. This is according to which large-scale human disasters in pre-industrial Europe and the Northern Hemisphere were directly caused by the economic recession caused by climatic change. Apart from the potential for human adaptation to climate change, the temporal unpredictability of the natural environment offers potential refugee fortes. The Younger Dryas was not as severe in Siberia as it was in the European segment of Northern Eurasia, and the Early Holocene warming during the Preboreal period (10.7 - 9.7 ka) was more in Siberia than in most of the Eastern Asian segment(Velichko et al., 2002). In the case of the Fertile Crescent, it appears that evidence is not always able to reveal regional climate variability. Based on data from archaeobotany, it was determined that the potential effects of the 4200 BP climatic variation on agricultural dynamics could not be broadly regarded as a catastrophic event. The Red Sea coastal plain, the Dhofar Mountains and surrounding littoral zone in Yemen and Oman, and the floodplain within the Persian Gulf basin were at least three core zones that functioned as population shelters in the Arabian Peninsula during environmental degradation/decline (Quaternary). Understanding how populations associated with these various refugia expanded and contracted will help us better understand the evolution of human populations in Arabia during the Pleistocene and Early Holocene. During times of aridity, these zones are the only reliable and offered freshwater in Arabia.

6.7 WETLANDS AND LAKES

Wetlands and lakes might be reliable archives/records of the local to regional climate. The age of lakes or peatlands determines the temporal scope of these archives. Because of glacial erosion, long sequences that predate the last glaciation are extremely uncommon in high-latitude environments. In low latitudes, sedimentation intervals during dry spells frequently restrict the amount of time records that can be preserved. As a result, records from lake sediments and peat are frequently brief and rarely span further than the current interglacial period. Furthermore, records from lakes and wetlands frequently show local conditions due to microclimates caused by topographical features and watershed characteristics, making it unable to make generalizations regarding climate conditions on a larger scale (both spatial and temporal). Nevertheless, a wide range of northern hemisphere lakes were examined and it yielded a lot of paleoclimate data. There are a lot of biogenic remnants in lake silt in addition to pollen grains and spores. Diatoms have been utilized to reconstruct several parameters, as they rely on the physicochemical characteristics of water.

Figure 6.3 Diatoms

Source: https://en.wikipedia.org/wiki/Diatom#/media/File:Diatoms-HCMR.jpg, *Creative Commons License.*

Diatoms (Figure 6.3) record summer productivity in high latitudes and their collections are mostly correlated with habitat parameters such as pH, nutrient availability, and water temperature. Diatoms proved to be helpful evidence to record salinity, paleolake levels, and hydrochemical conditions in low latitudes, where climate variations are characterized by abrupt changes in precipitations and hydrological regimes. Chironomids, which have been used to reconstruct temperature (records), are among the microfossils found in lake sediments. The study of the camoebians can be helpful in wetlands, such as peat bogs and marshes, where changes in the water table are closely linked to the hydrological budget. Other microfossils, such as ostracods, are found in terrestrial aquatic environments, but not enough is known about their distribution to allow for quantitative estimations of certain climate-related characteristics. Mesoamerica, which includes portions of the Southwest United States, Mexico, and Central America, is a large geographic region comprising deserts, semi-arid uplands, and tropical highlands and lowlands. In addition, Mesoamerica is home to many ethnic groups that have come together over time and space to share important characteristics such as their reliance on rainfall/precipitation, the pyramid temples and water mountains, staple foods like maize, and beans, and solar calendrical systems, their avoidance of hard metals, their lack of wheeled carts, the importance of ancestors, major deities and others. One other thing these societies had in common was how climate change affected these communities that relied on rainfall or precipitation.

Figure 6.4 Mayan civilization

Source: https://upload.wikimedia.org/wikipedia/commons/5/51/Chichen_Itza_3.jpg, *Creative Commons License.*

6.8 MAYANS-A CASE STUDY

As hunters and gatherers, the Mayans (Figure 6.4 is an example of their civilization) lived and adapted in the southern lowlands for millennia, starting more than 12,000 years ago(Kennett et al., 2022). With an increasing population, the Maya, who had been cultivating plants for thousands of years, became permanent farmers by 1000 BCE, residing in farms and small towns and dependent on cultivated beans and maize. Leaders had to arise to manage, settle conflicts, and distribute resources as the population grew. By the year 100 BCE, aristocrats who rose to become kings gathered their treasures in cities and constructed palaces, temples, marketplaces, and ultimately enormous reservoirs. As reservoirs dried up, the support for the ruling class eventually faded. This led to an urban migration in which farmers abandoned homes and many left the area in search of new places to support their families and survive. Early in the sixteenth century, the Spanish arrived in the northern lowlands, bringing with them cash crops and the first epidemic diseases that would fundamentally change the Maya region. Conquistador ancestors still have an impact today; for example, they control agricultural and sociopolitical structures.

Figure 6.5 Douglas fir

Source: https://en.wikipedia.org/wiki/Douglas_fir#/media/File:Pseudotsuga_menziesii_7971.JPG, *Creative Commons License.*

Figure 6.6 Montezuma bald cypress

Source: https://en.wikipedia.org/wiki/Taxodium_mucronatum#/media/File:El_Sabinal,_Salto_de_los_Salado,_Aguascalientes_51.JPG, *Creative Commons License.*

Mesoamerican ethnic and cultural groupings, dating back more than 13,000 years, are a reflection of the region's ecozone diversity(Turner & Klaus, 2020). People settled and farmed over time, giving rise to several ethnic groups with a shared history. Among the famous are the Olmec, who lived on the Mexican Gulf Coast between 1200 and 500 BCE, Teotihuacan, in central Mexico, the Zapotecs of Monte Alban, in Oaxaca, Mexico, 500 BCE to 900 CE, the Classic Maya, who lived in the southern lowlands of what is now southeast Mexico, northern Guatemala, and Belize, between 250 and 850 CE, the Toltecs of Tula, in central Mexico, between 950 and 1150 CE, the ancestral Pueblos of the Southwest United States starting in the late 1200s CE, and the Aztecs, who were centered in Tenochtitlan, Mexico, from 1428 until the Spanish Conquest in the 1520s.

6.9 RECORDS BASED ON TREES

With the help of Douglas fir (*Pseudotsuga menziesii*) (Figure 6.5) and Montezuma bald cypress (*Taxodium mucronatum*) trees, (Figure 6.6) researchers have been able to create precise rainfall histories/records through dendrochronology or tree-ring dating. These results can then be compared to sediment core and speleothem records from other parts of Mesoamerica.

Tree-ring studies demonstrate that the rise and collapse of Teotihuacan, Tula, Tenochtitlan, and other civilizations are preceded by modifications in rainfall patterns. The rise of these regional communities was associated with drought to pluvial transitions, suggesting that changing freshwater availability in the semi-arid Mexican highlands favored urban population growth and that forcing from the Pacific and Atlantic Oceans controlled this hydroclimatic change. Economies were financed by agricultural surplus; therefore, any disruption of crop cycles would destabilize the sociopolitical equilibrium.

The spread of infectious diseases, the ending of previous associations, and other circumstances combined with prolonged seasonal droughts brought an end to the Aztec empire. Remarkably, some researchers discovered pluvial maxima at approximately 1450 CE based on stalagmites from southwestern Mexico(Oster et al., 2019).

This is in contrast to dendrochronological data that indicates a drought that occurred just before the Famine of One Rabbit. However, an excessive amount of rain can harm crops in the same way as less rain can since it can wash away seeds, cause rot, attract more pests, and other problems. Droughts in particular have contributed significantly to other Mesoamerican disasters in the recent past and will continue to do so.

6.10 DROUGHT PATTERNS

Droughts or shifting patterns of rainfall have been a recurring issue in the histories of Teotihuacan, Tula, and other ancient Mesoamerican communities(Elliott, 2016). The northern border of Mesoamerica, which is home to ancestral Pueblo societies and has a long history of abandonment and resettlement beginning in the late 1200s CE matching with periods of severe droughts, has a similar pattern to what we see in the Southwest region of the United States. The tropics receive a lot of solar radiation all year round since they are located between 23.5° north and south of the equator. In comparison to the climates of non-tropical regions, the tropical zone between 0° and 10° latitude has high levels of radiation during the day and generally high temperatures with only relatively minor daily and seasonal fluctuations. Some of the intricate and varied ecosystems are made possible by these tropical conditions. The southern Maya lowlands are examples of humid semi-tropical environments with distinct wet and dry seasons. These regions are situated between roughly 10° and 23.5° latitude. Variations in water quality and availability affect a variety of environmental and socioeconomic aspects of life, such as the abundance of resources, the reproductive cycles of plants and animals, and the locations and lifestyles of individuals, including transportation, and so forth. Both the wet and dry season extremes have an impact; insufficient rain reduces water quality and supply, which causes crop failures and malnutrition. However, excessive rain results in flooding, low water quality, and famine.

In the past, communities that relied on rainfall were vulnerable to climate change. Extreme weather and its effects, such as seasonal rainy and dry seasons, tropical storms, and diseases transmitted by water, put communities at risk and prompted the development of locally adaptive measures. Nevertheless, these communities sustained themselves for millennia by utilizing large-scale water management systems in addition to small-scale subsistence technology, all of which were integrated through a low-density agrarian urban system that connected farms, communities, networks of exchange, and resources. Climate change was a significant aspect that influenced its history, as the Classic Maya of the southern lowlands serve as an example. Rainfall was essential to the ancient Maya society's ability to carry out agricultural tasks and sustain their economy. Mayan inscriptions and methods of interacting with the environment that are abundant in water indicate the significance of water. The extremes of water availability were a certainty that the Mayans often had to deal with. They framed strategies for storing and allocating water during the dry season and controlling and distributing it during the rainy season. The dry season, which lasts until May or June, comprises a three- to four-month stretch of completely dry weather during which humidity and temperatures increase. Farmers worked hard in their fields throughout the rainy season. They employed both large- and small-scale strategies to adjust to the five-month dry season. Large reservoir systems were built by them and incorporated into the urban design. The maintenance of small areas, which are rain-fed natural depressions that would have eventually dried up as the dry season progressed, and the collection and storage of water in jars and other containers were examples of small-scale and local adaptations. The ability of the Maya to forecast the termination of the dry season and the start of the rainy season was crucial, even concerns about the quantity and quality of water. Farmers have to sow as soon as possible before the rainy season to have the highest harvests of maize, beans, and squash. Specifically, maize needs a lot of water. In tropical regions, precipitation patterns are already unpredictable in terms of both quantity and time. In the Maya lowlands, the onset of the rainy season can fluctuate by as much as five months, and annual precipitation can vary from 1350 to 3700 mm(Lucero et al., 2011). Because of the seasonal variations and the absence of extensive irrigation infrastructure, seeds would decay in the case of late rains and would not germinate in the case of early rainfall.

6.11 WATER MANAGEMENT IN THE MAYAN ERA

Water management systems were built because people needed access to clean, reliable, and abundant water. The Mayans constructed water management systems before starting any building. The earliest of these systems in the southern lowlands is supported by evidence from northern Belize, where the Maya dug shallow ditches draining the margin of swamps to control water flow around 1000 BCE. The complexity of water systems increased, and by the Late Preclassic era, they comprised micro-watershed systems where the Maya took advantage of the natural landscape. These low-lying, gravity-driven water systems would have seen more material/soil seeping into them as erosion grew as a result of the increased forest removal. During 250–550 CE, reservoirs grew in significance and complexity(Gibling, 2018). More people moved into highland areas with relatively large plots of land but limited surface water as the lowland population grew, necessitating the construction of reservoirs. Throughout the Late Classic era (c. 550–850 CE), the Maya depended more and more on these reservoirs as their size increased. Large-scale initiatives were necessary for the creation of monumental architecture, especially during the Classic period. Massive efforts in mining were needed for the construction of gigantic pyramids and palaces. The elevated micro-watershed systems, which included reservoirs, dams, canals, sluices, filtration systems, and systems for the collection, storage, and distribution of water, are the best examples of constructed reservoir systems. These reservoirs were filled during the rainy season and were accessible for most of the dry season due to the use of the landscape to restrict the flow of water into them. Water was distributed to lower-lying settlements via a gravity release system throughout the dry season, particularly in the early months when the water quality was adequate. But as the dry season went on, the quality of the water declined, and a large portion of the water was redirected to ponds and agricultural fields. Water that was reliable, abundant, and clean was essential to this entire system.

A large portion of the rainfall was naturally purified by the karstic terrain of the southern lowlands, as well as other natural processes and landscape elements. However, it would have been difficult to maintain water quality in large reservoirs over the expanded dry season. Reservoirs were built adjacent to buildings, which increased the risk of human waste pollution due to the proximity and accessibility of these areas. Other purification techniques were used because the karstic topography by itself was ineffective in preserving pure water in these reservoir systems. The Maya used their extensive understanding of the environment to establish a wetland biosphere that filtered the water in the reservoir. This biosphere was made up of a balance of hydrophytic and macrophytic plants as well as other organisms. Since water lilies (*Nymphaea ampla*) are delicate hydrophytic plants that can only thrive in calm, clear, shallow (1–3 meter) water that is not too acidic and does not have too much algae or calcium, their presence on reservoir surfaces is indicative of clean water. Water lilies also show that clean waters are made possible by the sediments in which they are rooted; if there is an excessive amount of organic debris, the gases emitted during this process create a poisonous environment that is unfavorable to water lilies. Furthermore, the Maya produced a more neutral or slightly alkaline environment, which many Nymphaea species prefer, by surrounding the reservoirs with clay to stop calcium from seeping into the waters. Reservoirs have also employed alternative methods of water filtration, such as the use of gravel and sand. Reservoirs were not only used to store water; aquatic plant and animal life from the pools was used as a source of food, medicine (medicinal plants), and other uses. The bottom sediments could also be used as fertilizer.

The king's primary responsibility was managing water resources by planning the construction and upkeep of reservoir systems and by praying to gods and ancestors to guarantee an abundance of rain. Kings kept their subjects, along with their labor, services, and goods as long as they fulfilled their duties as water managers. Although the connection between social transformation and climate change has been extensively studied, it is still a complicated link that requires further investigation. Although the climate has often been identified as an important causative factor to the decline of the Classic Maya, this relationship is also complex because millions of people had to cope with centuries of continuous drought, and there were hundreds of Classic Maya cities, each with a distinct and lengthy history. The fact that millions of Maya people still live today shows that, despite the breakdown of their standards, they survived. There is more and more precise evidence of the extended droughts. Speleothem data from caves in west-central Belize and northwest Yucatán, Mexico, show that the region had multiple droughts between around 800 and 930 CE(Wu et al., 2019). The water supply in reservoir systems was severely disrupted by these droughts, which also made problems like population increase, resource misuse, and erosion worse for the Maya civilization. After the Late Classic period (c. 550–850 CE), when Maya population densities were at their peak and people required access to clean water, droughts occurred. The intervals between the peaks were brief but insufficient to allow recovery. Depending on the specific social, environmental, and political conditions, the droughts had varying effects on different centers at different times.

6.12 SOCIETAL CHANGES

Substantial societal changes were brought about by the collapsing political and social system. Farmers stopped supporting the Maya emperors in the late 800 CE because they could no longer supply enough water for everyday needs. 90% of people moved to coastal regions or along major rivers from the interior southern lowlands, which were home to the majority of urban centers after the monarchs lost power as a result of reservoir systems failing. There were instances where there were no water management systems. These centers were self-sufficient because they were situated on fertile alluvium and had access to water from the river and nearby aquifers. Since these areas were still inhabited until the Spanish conquest in the early 1500s, the droughts did not have the same effect on them as they did on other lowland areas. However, the Maya urban centers collapsed elsewhere in the lowlands, and the population moved permanently to coastal and riverine habitats. People who left urban areas moved in different directions. Even though the reservoirs provided water to hundreds of thousands of people for nearly a millennium, the political system fell apart when the rains stopped and the royal rituals that had earlier been used to prove a king's authority also failed.

Crops, forests, water supplies, human health, and fertility would have all been impacted by multiple droughts. Even though the majority of people left the inner lowlands, individuals who persisted formed communities and kept using the unmaintained reservoirs. The Mayans lived in small towns along rivers and lakes in the Petén, the Belize River valley, the northern Belizean wetland areas, and the coastal regions of Mexico, Belize, and Guatemala. For instance, the Maya inhabited the Petén Itza lakes region of Guatemala until the invasion and the Postclassic period along the Belize River. During the Postclassic, small communities continued to exist. People moved to coastal regions and areas close to large rivers because they offered new economic opportunities and water supplies. Market towns, interior trade, and marine trade quickly developed in these areas. There is proof of the emergence of these new Postclassic settlements at Bacalar, Cozumel, Chetumal, and Chichén Iztá(Lucero & Larmon, 2018). Maybe because they prospered in their new social and environmental environments, the Maya never went back to the southern lowland regions. Research on the Sahara and North Africa in general has long recognized the role that the environment plays as the primary catalyst for social change. The environment was significant, and it even appeared to control change through processes of adaptation.

6.13 THE SHIFTS AND PHASES

During the Late Pleistocene, the Sahara experienced periodic cycles of wet phases that were more conducive to human occupation and cycles of arid phases. These cycles were controlled by the precession of the equinoxes, which determine summer radiation levels over the Earth's surface and the relative advance and retreat of glaciers. The Last Glacial Maximum (LGM), when ice in the Northern Hemisphere prevented the seasonal shift of the Inter Tropical Convergence Zone (ITCZ), to the north, which determines the beginning and end of the monsoon season in the tropics and subtropics, left the Sahara dry. The subtropical anticyclones, which prevent monsoonal rainfall, are strengthened by northern cold currents that reach southern latitudes. As a result of the cold weather, the monsoon is unable to return northward every summer. Moreover, on the western side of North Africa, the polar anticyclones obstructed the cyclonic rainfall originating from the Atlantic Ocean. The northern slopes of the Ahaggar peaks featured snowfields, and pollen data suggested that temperatures in Algeria, northern Mali, Niger, and southern Libya dropped as much as 6°C. The environment gradually deteriorated following the last interglacial, when Saharan territories benefited from sufficient rainfall that also led to the creation of lakes. The Nile Valley and the coast of North Africa did not see impacts that were as severe as those experienced by the continental areas. The path of the Nile River was extremely slow due to the heavy sediment load, which resulted from the diminished activity of the White Nile and Blue Nile. The Nile Valley was repeatedly invaded by dunes from the Western Desert, and the fresh lakes that resulted from these dams provided the perfect environment for human habitation. A harsh period occurred in the Sahara during MIS 4 (80 ka BP - 60 ka BP), and partially during MIS 3 (60 ka BP - 30 ka BP). It is believed that migrations to the north may have resulted from the exceptionally dry conditions of the tropical regions in the phases that followed MIS 5. But without comparable data repositories, reconstructing a consistent paleoclimatic framework over such a vast region is challenging. The paleo-lakes region of Bir Tarfawi and Bir Sahara, located in the southernmost portion of the Egyptian Desert, provides an essential data repository for the Eastern Sahara(Barich, 2018). Based on nine SAR-OSL dates (Single Aliquot Regeneration Optically Stimulated Luminescence), improved geochronology of the Middle Stone Age lakes allowed for the establishment of inter-basinal correlations as well as the specification of the expansion phases of the Pleistocene paleo-lakes that were supported by groundwater at Bir Tarfawi and in the nearby area of Bir Sahara.

6.14 SEDIMENTS AND DEPOSIT RECORDS

There are also descriptions of extensive lacustrine sediments from Dakhla Oasis that are deposited in basins of perennial waters as marls (authigenic calcite silts). The MSA evidence stratigraphically linked to the lake deposits further suggest that these paleo-lakes were active primarily during MIS5. The paleo-lakes found in the Oasis of Kharga south of Dakhla can also be linked to the same MIS5, which is unquestionably the wettest phase in the Egyptian Desert. The major phases of these lakes have been established by Uranium-series datings on tufas from distinct regions of the oasis. These updated dates place the lake phases between late MIS6 and MIS5 (127.9 ± 1.3 ka - 104 ± 14 ka). The tufas of Wadi Midauwara, have produced an intermediate period of roughly 133 ka, but it seems to be marginally more recent than 124 ka. Lakes are not known to exist in the Eastern Sahara before MIS1 or after MIS5. The extremely low monsoonal indexes found between MIS4 and MIS2 imply that any occasional precipitation may have only been brief and of low intensity. Some evidence of these intermittent rains has been dated to approximately 50 ka in the Kharga region, while lake deposits in the Dakhla region have been dated to 40 ± 10 ka and 62 ka respectively. Water availability and associated population declines occurred throughout the Sahara during MIS2 (25–15 ka BP), particularly during the LGM and Younger Dryas (YD) periods. The aeolian MIS2 deposits in the Jebel Gharbi region of northwest Libya are a definite sign of the northern boundary of the Sahara expansion towards the north and the Mediterranean. Only the Mediterranean coast and the Nile Valley have records of human settlement from this era. Starting at 15 ka BP and interrupted with the YD, the re-establishment of the monsoon regime with the shift towards the north of the ITCZ was then resumed by 10,000 BP(Revel et al., 2010).

There were noticeable changes in the social groups throughout the Pleistocene/Holocene transition. Even the Saharan regions, which had been abandoned for a long time, were gradually reoccupied by tribes that tended to expand their resource catchment as the monsoon cycle resumed. Thus, the Nile Valley (Figure 6.7) and the Eastern Sahara comprised an ecosystem of hunting activities with the goal of widespread resource exploitation.

Figure 6.7 Nile Valley

Source: https://en.wikipedia.org/wiki/Blue_Nile#/media/File:ET_Bahir_Dar_asv2018-02_img32_view_from_Bezawit.jpg, *Creative Commons License.*

Populations gained extensive knowledge of their environments, which they then heavily exploited, thereby expanding the base of exploitation. People were not compelled to make long migrations because they utilized resources that were readily available or those that were close to their homes. When environmental production was higher, there was less nomadism and more domestic or semi-residential attitudes developed. Based on these presumptions, it was proposed that these groups might represent an environment that has already been domesticated. Most of Egypt and northern Sudan experienced wet climates during the Holocene. Different transient lakes (playas) filled the basins and left the lacustrine sediments that now produce an abundance of remnants from the old settlements that were formerly located on the beaches, including tools, hearths, and animal skeletons. Conversely, dune formations, which were mostly observed during 9.7, 9.2, and 8.2 ka cal BP, suggest that there were powerful climatic events that led to the abandonment of the inhabited sites(Srivastava et al., 2020). Since the Early Stone Age, people have lived in the Egyptian Western Desert, which serves as a buffer zone between the Nile Valley and the Central Sahara. However, the most important settlement phase began in the Early Holocene (8000–6000 cal BC), and was marked by fluctuations in the wet and dry weather. The vast playa deposits found in the Gilf Kebir, the Nabta region, and the Farafra and Dakhla Oases collectively served as the foundation for reconstructing the paleoclimate and human occupation of the Egyptian Western Desert. Compared to earlier studies, research conducted in the 1990s in the southern region of Nabta has made it possible to gain more knowledge. El Adam, between 10.8 and 10.0–9.9 ka cal BP; El Ghorab, 9.5–9.2 ka BP; and El Nabta/Al Jerar, 8.9–8.1 ka BP, are the three primary wet episodes of the Early Holocene. The three phases are divided by roughly a century of dry intervals, which means that they were formed in a time period that is very valuable on a human scale and that resulted in a noticeable restructuring within the populations. Situated in the heart of limestone depressions, the five oases of Siwa, Bahariya, Farafra, Dakhla, and Kharga are situated further north. The deepest limestone basins, which are actual water reservoirs, appear to have been created by karstic action.

6.15 CULTIVATION AND IRRIGATION

Specifically, the early instances of proto-cultivation that have been identified in these regions have made it possible to reinterpret the agricultural contexts of the Nile Valley, some of which are comparable to the Near Eastern Neolithic period. According to the most recent studies conducted in the Western Desert, winter rains would have been brought about by a descent of humid air commencing around 8.1 ka cal BP, at least as far south as Dakhla. As a result, human populations were able to live there all year, providing a far more stable model than was previously thought. By opening a few wells to retain water for the driest months, the southernmost section of Nabta achieved the same outcome. As a result, the populations may stay at their home, which are in the basins' deepest sections, from late fall until the summer rains arrive. They could then go back to hunting. When the edible plants reached maturity in the early winter, they were preserved in protected vessels so they could be used in the hardest times. The growing significance of vegetable-based foods in the diet ultimately forced an alteration to the annual cycle. Communities were forced to divide into two groups: those tasked with caring for and gathering plants would have stayed at the residential base during the whole fall and winter season; those tasked with grazing the herds, on the other hand, would have walked nearby, mostly on the surrounding plateaus. Arabian human populations have followed a unique course during the Holocene, which is very different from the paths taken by populations in Mesopotamia. While farming communities grew throughout the Fertile Crescent, which stretches from the Levant to the Taurus and Zagros mountains, during the first half of the Holocene, agriculture was not adopted until much later by Arabian populations. This is likely due to the Late Chalcolithic period in Northwest Arabia and the Bronze Age beginning in Southwest Arabia and Southeast Arabia. Prior to that time, between the sixth and fourth millennia B.C., nomadic hunters inhabited the Arabian Peninsula. These people lived in a savanna-like habitat that was made possible by the

Holocene phase, where they hunted and herded sheep, goats, and cattle. Groups that were experts in fishing, collecting shells, and using mangroves were found around the eastern shores.

The most valuable resource for humanity is water, which is used for agriculture in arid areas as well as for personal consumption. An important turning point in the history of water management is the Neolithic(Davison et al., 2006). People needed to rely on water more than ever. This resulted from the growing population, sedentary/inactivity, demographic expansion, and agricultural advancement. It became important to obtain and preserve water as soon as people began to live there in the Neolithic era. In the settlements next to lakes, rivers, and springs, potters might be used to transport and store water. However, when surface water supplies were only momentarily available, people had to adopt hydraulic methods for either extracting them or using groundwater. The oldest wells date to the 9th or 8th millennium B.C. and were discovered in Pre-Pottery Neolithic settlements in Cyprus and the Levant. Given that they were discovered in the interior of villages, these shelters were likely utilized for food, both human and animal. About Arabia, the northern border of the peninsula, in the Jordan desert, yielded the earliest indications of hydraulic activity. The northern portion of the Arabian Peninsula, near the Nafud, is where the earliest evidence of water management in the region dates back to the Neolithic era. Rainfall must have been more consistent back then, and the denser vegetation encouraged slower or less frequent surface runoff as well as increased recharge of the groundwater. Therefore, the crossing of dry regions was made possible by the abundance of watering spots (lakes and aquifers with high water tables). Standing stones are an important part of the unique residential, ritual, and burial architectural legacy that was built by nomadic pastoralist cultures in the Levant during the 5th century B.C. These pastoralist communities were skilled in hydraulics because they dug wells next to lakes, often in conjunction with troughs, to supply water to their herd. They also used small dams to capture runoff water.

6.16 EXCAVATIONS AND FINDINGS

The well/watering complex D15 from Qulbān Banī Murra has been dated to the second half of the 5th century B.C. (Charbonnier, 2018). There have been other such structures identified at the location, but D15 is the only one that has been excavated. This massive construction is made up of a well with walls consisting of vertical slabs. The well's diameter is approximately 1.2 meters, and it was excavated through wādī gravels. Its depth is unclear because, for safety considerations, excavation had to be halted about 4 meters below the surface. Given that the complex D15 was situated in an area where the main aquifer was pushed up due to a sub-surface narrowing of the bedrock and received additional influx from nearby aquifers, it appears that populations living in the fifth century B.C. had a thorough understanding of groundwater. In Saudi Arabia, wells connected to troughs have been found. The site was inhabited from the Late Neolithic to the Early Bronze Age. The site is marked by a series of endorheic depressions that trap water and sediments, and there are indications of dwellings. A well connected to a rectangular trough marked by a series of standing stone slabs was discovered during excavations. Despite its poor preservation, the well bears a striking resemblance to the one of complex D15 in Qulbān Banī Murra due to its architecture, size, and the masonry in its top section. The archaeologists propose a dating of the fifth century B.C. based on this data. Up to five meters below the surface, the well was recently partially dug. A semi-circular row of standing stone slabs encircles the well from the excavation, which is approximately 0.8 by 0.5 meters. The trough that it is connected to is approximately 2 by 0.6 meters. Since the site investigation only recently began, more information and improved dating should become available in the years to come. According to recent surveys and archaeohydrological research, the majority of Rasif wells were situated in endorheic depressions and most likely tapped a shallow aquifer that was filled with sediments from flooding. Some of these shallow wells were likely abandoned after the Chalcolithic because of the depletion of the shallow aquifer as a result of precipitation loss and aridification, while others were deepened to access the underlying fossil aquifer.

The aridification that occurred after the fifth century B.C. likely made groundwater extraction more challenging, and the researchers (at Rasif region) have shown that endorheic depressions may be improved by creating dams that held runoffs in large pools. This runoff water collection maneuver appears to have been upgraded and used up until recently. The Bedouins might have utilized it to store water and maybe to establish transient farming areas following periods of precipitation. The research work in Rasif demonstrates how human communities adjusted to the mid-Holocene water scarcity and climate aridification. During the 4th century B.C., a decrease in water resources caused the population to settle in more resource-dense areas and create sedentary oases. However, some groups continued to be nomadic pastoralists. Some researchers postulated that the indigenous people created these oases by utilizing their knowledge of managing water. The Chalcolithic and Bronze Ages are less popular in Arabia, although they coincide with the fast growth of city-states and urban centers and the hydraulic systems in the Near East and Mesopotamia. Depending on the location, Arabia's Bronze Age spans from the end of the fourth/beginning of the third to the end of the second century B.C. During this time, communities began to adopt agriculture, and permanent villages became the standard in Arabia (neolithization). Throughout this time, Southern Arabia also produced its ceramics. The question of whether these changes resulted from direct interactions with agricultural cultures in the north or from internal factors, that is, from local communities adapting agriculture to suit new demands and the evolution of the social structure has been challenged. Though we can only speculate, we can assume that it was a combination of the two. In Arabia, very few water systems from the Bronze Age are known to exist, and many of the hydraulic buildings that are thought to date from this era are incorrect. Moreover, absolute dating isn't accessible all the time. There is little information available for the 2nd century B.C., although some data is available from the end of the 4th/3rd millennium B.C. Some projects follow the question of how social and economic life, and related water management and land use, adapted to disappearing steppes, lakes, and lowering of water tables after 4200/4000 BCE(Gebel & Wellbrock, 2018). As there is currently little evidence in this emerging field of study, research is conducted using a set of testable hypotheses that are continuously refined by fresh findings from laboratory and field research.

6.17 ARID ENVIRONMENTS

It may take more time and effort to complete the reconstructions in the desert lands than in other areas because they hardly ever provide intact sedimentary environments and organic debris. The Mid-Holocene development of human activities, water use, and connection with nature in Arabian arid water-deficit regions followed different conditions and frameworks than the stratified hydraulic cultures of ancient Mesopotamia and Egypt. The vulnerability patterns had more to do with structures implementing intricate adaptive systems to maintain social and economic equilibrium. Examples of such strategies included widespread land use reorganization, organized territorial conflict and warfare, including forced migration. Naturally, these socioeconomic frameworks are the direct outcome of the many adaptive techniques that are employed in arid regions to deal with the varying patterns of vulnerability. In moderate situations, even periods with recurrent hazards can influence human behavior and cultural development. The potential of socioeconomic adaptations in complex productive systems in moderate zones should not be undervalued, even though this statement serves as a crucial reminder that climate-deterministic approaches should not be disregarded because of cultural deterministic approaches typical of the humanities. This was particularly true of the ancient hydraulic civilizations that grew up beside the major rivers. However, climatic determinism must predominate in the research in areas that are easily susceptible to drying. Throughout times of stability and instability, several forms of vulnerability constantly interact. In essence, a vulnerability system is made up of social, environmental, and ideological weaknesses. In general, the evidence suggests that had drier conditions not suddenly forced them to adapt to sedentary socioeconomics starting around 4200/4000 BCE, mobile pastoral societies would have continued their conservative lifestyles, water management, and belief systems. The primary driver of change in these societies was climate forcing as soon as it began to shift. In the past, climate forcing acted as the primary driver of change in the Arabian desert. The steppe pastoralists of the Arabian Mid-Holocene could only respond to the constraints of their surrounding environments; similar responses may be found in the oasis communities that followed, which exhibit comparable conservative characteristics as well as comparable cultural and socioeconomic expressions/indications. Climate change had a less direct effect once alternative risk management strategies, such as socioeconomic and innovation strategies, including high-level aggression of the stratified complex societies in the modern Fertile Crescent and Nile Valley were available.

In addition to identifying sparsely preserved pockets of sedimentary environments and installations, dry-land archaeology/archaeohydrological and land-use research in Arabian environments pose challenges in tracking related network elements and catchments due to the geographical extensions of the pre-historic/historical landscape cultures. The habitats of arid Arabia were always the most sensitive to even slight variations in the climate, which had significant effects on land use, population dynamics, water management, and therefore cultural advancements. Because there were few ecosystems and therefore no unsustainable adaptations or innovations possible, deterioration through aridification did not lead to significant unsustainable developments until recently. Instead, the harsh environmental conditions necessitated sustainable adaptation. We scarcely have pertinent and consistent data on climate developments for Northern Arabia during the mid-Holocene (Lüning & Vahrenholt, 2019). We may thus rely on supra-regional data for research studies, including the comparison with Early Holocene conditions. The Early Holocene was characterized by more humid conditions due to the shift of monsoonal air masses towards the north, as shown by paleoclimatic and paleoenvironmental studies conducted in the Arabian Peninsula, as well as in the Western and Eastern Saharan region. Although there is evidence of climatic deterioration with increasing aridity towards the mid-Holocene, these deteriorations correlate with global Rapid Climate Changes (RCCs). The Early Holocene environments that characterize the hyperarid regions of today were steppe environments. Although lacustrine and steppe environments are well documented, the Arabian Peninsula remained relatively humid during the 6th and early 5th centuries. This served as the foundation for the previously described mobile pastoralism, which had vast networks. The moist climate of the 6th to early 5th century led to denser vegetation in terms of hydrology and vegetation. This was sustained by many positively correlated factors, including decreased or slowed surface runoff, decreased sediment transport, increased groundwater recharge that resulted in filled aquifers, and reduced sediment transport that promoted soil formation. Rainfall must have been more reliable, resulting in more regular and permanent water sources rather than sporadic ones.

Chapter 7

AQUACULTURE AND CLIMATE CHANGE

Contents

7.1 INTRODUCTION

The Sustainable Development Goals (Figure 7.1) represent a pressing need for rich and poor countries to work together in a global partnership. It takes coordinated efforts to address climate change, protect our seas and forests, enhance health and education, lessen inequality, and boost economic growth to eliminate poverty and other forms of deficiency.

Figure 7.1 Sustainable Development Goals

Source: https://its-wiki.no/images/f/fa/Visjon2030_overview.png, *Creative Commons License.*

The SDGs also include targets that can be met by boosting the management and contribution of aquaculture to food security and nutrition, particularly in terms of natural resource utilization. This is in addition to the target of nourishing the 10 billion population by the year 2050. The food industry with the quickest rate of growth is still aquaculture, which is predicted to offer enough aquatic food to meet demands. An extensive analysis of global aquaculture up to 2018 is provided by the Food and Agriculture Organization of the United Nations in its most recent biennial assessment. Since the late 1980s, the capture fisheries have been inactive, and aquaculture has played a significant role in boosting the supply of fish available for human consumption.

7.2 FISH CONSUMPTION AND FUTURE

Fish and fish products exported in 2019 accounted for around US$162 billion, or about 37% of the total fish production(Nunes et al., 2019). Compared to low-developed countries, the inhabitants of rich countries eat more fish per person. The average annual consumption of fish per person rose gradually from 17.4 kg in 1961 to 26.4 kg in 2007, but then leveled off at 24.4 kg from 2007 to 2017.

The annual fish consumption rate in developing countries climbed drastically by 2.4% annually between 1961 and 2017 (5.2 to 19.4 kg per capita. The consumption of fish in low-income food-deficit countries (LIFDCs) increased from 4.0 to 9.3 kg per capita between 1961 and 2017. The least-developed countries (LDCs) in Africa have had a modest rate of growth, with an average annual increase of 1.3%, from 6.1 to 12.6 kg per capita. However, it is predicted that per capita fish consumption in Africa will decrease from 10.0 kg to 9.8 kg between 2018 and 2030, at a rate of 0.2% per year. The World Bank predicted that fish consumption in Africa could drop dramatically over the next 20 years, despite expected increases in fish consumption in Asia and Latin America. Such a prediction is alarming and deserves consideration at a time when the crucial role that fish play in ensuring the security of the world's food supply and nutrition has been widely recognized.

Aquaculture is the production of fish using a range of farming systems in freshwater, brackish water, and saltwater using extensive, semi-intensive, and/or intensive production techniques(Araujo et al., 2022). Production systems span from complex, high-tech industrial systems to small-scale, low-tech processes. Aquaculture has become accustomed to sustainable intensification as a result of multiple sectors fighting for few primary resources, which include both land and water. Aquaculture prosperity is largely dependent on increased productivity, or output per unit of land, water, and/or energy. This is evident from the increasing technological advancement and modernization. Over 70% of aquaculture is still produced by smallholders utilizing ponds, cages, pens, or raceways, depending on the species farmed and the availability of land and water. This is even though aquaculture is being intensified and modernized through technological developments. Marine aquaculture offers the chance to increase seafood production worldwide given the growing demand for animal protein and the finite capacity of wild capture fisheries. It is clear that providing food for 10 billion people by 2050 would be a difficult undertaking with several obstacles in obtaining the necessary resource base. Aquaculture will grow more common in rivers, lakes, reservoirs, and the ocean in the upcoming decades as land-based aquaculture will continue to compete with other land uses. In Asia, it's common practice to raise freshwater fish in cages in lakes, reservoirs, and rivers as well as in earthen ponds.

7.3 FISH FARMING

Fish farming in cages dates back to the 1800s in Southeast Asia when freshwater fish were first kept in traditional wooden-framed cages in the Mekong Delta (Cambodia). Since the 1950s, when large-scale marine fish farming began to appear in Europe and Japan, cage culture has grown significantly and increased.

It is currently developing quickly in response to the difficulties brought on by growing demand and globalization. Intensive cage culture has spread in inland, coastal, and marine (both nearshore or offshore) habitats as a result of the lack of suitable sites. Attempts were made to group together already-existing cages to use more intense cage-farming methods.

***Figure** 7.2* Fish farming

Source: https://en.m.wikipedia.org/wiki/File:Cage_fish_farming_in_lake.jpg, *Creative Commons License.*

Fish farming (Figure 7.2) in cages offers a lot of benefits over conventional pond culture. It is advantageous for optimizing open water resources (productivity), sharing water resources for various purposes, creating a range of farming techniques, preserving ideal water quality conditions, enhancing health management, preventing or eliminating predation, assuring high stock survival, reducing initial investment, and making harvesting simpler. When scaled suitably, cage-based aquaculture offers lower operating costs per unit of production than other intensive aquaculture systems. Because cages placed in places with sufficient water flow velocity require no energy for natural water exchange, the carbon footprint of cage culture is also low due to the low investment cost per cubic meter of farm capacity.

The FAO in 2007 reported that 62 countries produced cage-cultured species in 2005(Tidwell, 2012). Nearly all aquaculture of marine finfish takes place in cages. In Asia, a wide variety of freshwater fish species are also raised/grown in cages. Cage cultures are built and run using locally accessible resources and basic technology in many countries, taking into account the local social, economic, and environmental situations. The present spectrum of cage-culture systems includes large-scale commercial cages used for salmon and other marine species including sea bass, sea bream, and tuna, as well as traditional small cages owned and controlled by families that are characteristic in most Southeast Asian countries. Cage culture has been used for a long time in Cambodia, however, the productivity from aquaculture is much lesser than that from capture fisheries. It is still the fastest-growing segment of the food production industry, rising by more than 18% between 2002 and 2014, and is dominated by small and marginal farmers in freshwater, brackish water, and marine environments. Marine cage culture in floating net cages located in estuaries and coastal seas is growing in Cambodia despite declining profit margins. Despite making up only 1.2% of Cambodia's aquaculture production, the marine cage industry was valued at US$7 million, or 6.1% of the country's overall aquaculture production. Finfish, including the Asian sea bass (*Lates calcarifer*), shellfish, including the green mussel (*Perna viridis*), and blood cockles (*Tegillarca spp.*)(Kibenge, 2022). One of the most often used methods for raising freshwater fish in the Mekong Delta is tilapia cages. It makes a major contribution to rural livelihood and food security. Nonetheless, diseases, fish mortality, and declining water quality have been problems for freshwater cage farms. Cage farming for tilapia and striped catfish, *Pangasius* spp., has developed. As a result, there has been a significant increase in farming intensity, especially in stocking densities. The number of cages has rapidly increased in many areas of the Mekong Delta. Even in areas with extensive industrial pollution, cages were set up, and production levels dramatically dropped due to the lack of sound management techniques. However, potential new sustainable cage-culture methods exist, primarily due to requirements for standards and certification. In addition to being mostly found in rivers, cage culture is becoming more and more common in many freshwater lakes across the globe.

7.4 EXPANSION

Cage culture has expanded in many reservoirs in northern Vietnam as a result of new government initiatives backed by financing, strategic planning, and law. This quick expansion is also attributed to the belief that fish raised in cages are of higher quality than fish produced in ponds, meeting export specifications and taste better. While there are many other ways to produce tilapia, one of the most widely used methods is cage cultivation in open seas. Since the 1950s, when it was first developing, there have been no environmental or regulatory limitations on the use of open-water resources, and tilapia cage culture has thrived in the majority of rivers. Cage farming was permitted in several of the suitable reservoirs. However, the number of cages in most farming areas has decreased due to the expansion of river cage culture, related environmental difficulties, and regulatory constraints. Many freshwater lakes in Africa are also extensively used for cage culture, primarily for the production of Nile tilapia. Numerous cage culture operations in African lakes, including Lake Volta (Ghana), Lake Victoria (Uganda and Kenya), Lake Malawi (Malawi), and Lake Kariba (Zambia), were reported. There have been reports of large-scale commercial tilapia cage farming operations in Lake Kariba, Zimbabwe. There were only 185 cages in 2006; these were located on Lake Volta (Ghana), Lake Malawi, Lake Victoria (Kenya), Lake Victoria (Uganda), Lake Kariba (Zimbabwe), and 30 cages on the same lake in Zambia. There are already over 20,000 cages in inland waters, demonstrating the continued growth of cage culture in Africa. Industrial marine fish farming has grown throughout several countries, presently producing 6.6 million tonnes of fish/year, the majority of which originate from sea cages. The first marine/sea cage farms were constructed in the 1960s in Norway and later in Scotland to raise Atlantic salmon, Salmo salar, using nylon trawl nets and a timber or polyethene framework. More commercial farms were added in the 1980s. Recently, more complex constructions have gained popularity, such as net cages made of different-sized nets and mesh. These create the ready-to-install cage constructions, which are made from high-density polyethylene (HDPE) or stainless steel. In Australia and Indonesia, barramundi and tuna aquaculture, respectively, use circular HDPE cages. Compared to farming inland, which is primarily focused on producing food and a sustainable rural livelihood, marine finfish aquaculture uses more sophisticated technologies and produces high-value fish at a higher profit. For this reason, efforts involving marine cage culture usually draw larger funding than those involving inland cage culture.

7.5 IMPACT OF THE WEATHER AND OTHER VARIABLES

While weather refers to the state of the atmosphere at a certain time and place, climate refers to long-term weather patterns that define a place. The more a meteorological variable deviates from its usual fluctuation range, the more it approaches an extreme, whether it be a temperature spike, a hydrological event, or an abnormality in air circulation. Climate variability can result from both anthropogenic forcing, like the production of greenhouse gases (GHGs), and natural processes, including large variations in ocean currents, volcanic eruptions, etc., inside the climate system. Many studies have been conducted on how aquaculture is affected by climate change. Some researchers assessed how aquaculture productivity was impacted by climate change and what it meant for adaptation, mitigation, and sustainability. Aquaculture is impacted by climate change in both direct and indirect ways(Cubillo et al., 2021). The physiology of the fish stock is affected directly, but changes in primary and secondary productivity, ecosystem structure, input costs or product prices, fishmeal and fish oil prices, and other costs associated with goods and services needed by fishermen and aquaculture producers can all have indirect effects. Production from aquaculture is entwined with other forms of food production. Because climatic changes are the primary factor affecting the food production system, studies have shown the detrimental effects of sudden climate changes on aquaculture and agriculture.

The majority of livestock production and some wild capture fisheries have lower greenhouse gas emissions per unit of protein generated by aquaculture. According to estimates, aquaculture was the source of 245 million tonnes of CO_2 equivalent in 2017 or roughly 0.49% of all human greenhouse gas emissions worldwide. This is significantly less than the carbon footprint associated with farming on land. Despite reports that mariculture has a lower greenhouse gas footprint than other food production industries, there are significant variations in the average greenhouse gas emissions footprints of the finfish, bivalve, and seaweed sectors as well as significant variability within sectors. This variance supports the need to find and put into practice techniques that increase mariculture's capacity to be climate-friendly, regardless of its current GHG emissions profile. It also supports the urgent need to rapidly cut GHG emissions on a global scale. Transport-related greenhouse gas emissions could rise significantly if the import of seafood becomes more profitable in developing countries. In addition to potentially reducing GHG emissions, shortening supply chains and developing regional markets may also increase food security and industry resilience during difficult times. Sustainability in aquaculture requires tackling the effects of climate change. Rising temperatures, diseases, harmful algal blooms, altered rainfall and precipitation patterns, sea level rise, uncertainty regarding supplies, altered sea-surface salinity, and extreme weather events are the main aspects of climate change that pose a threat to aquaculture production and sustainability. It can be difficult to forecast how each of these variables will affect aquaculture production because they might function separately depending on many variables like the season, the location, and the size of the particular sector. Throughout important phases of a fish's life cycle, the influence of these factors might also be apparent. Moreover, there is variability in how these factors affect aquaculture production.

7.6 SEA SALINITY

Sea salinity fluctuates as a result of increased evaporation brought on by warming temperatures and modifications to ocean circulation, as well as the direct effects of climate change(Du et al., 2019). Every aquatic organism has a salinity tolerance range that allows them to survive; any changes could result in death and decreased productivity. Fish that experience salinity levels over their normal tolerance are more vulnerable to infections due to decreased immunity, growth, and survival. Variations in the salinity of water often cause several species to die more frequently, which could have an impact on the social and economic viability by increasing the number of species lost and raising the cost of management. Research on the effects of salt levels below ideal requirements for specific species of shellfish and finfish is needed. Furthermore, little is known about how many commercially significant species react to variations in salinity brought on by climate change. Since salinity variations may encourage the production of resistant species, this knowledge is particularly helpful for aquaculture adaptation. Variations in precipitation patterns can result in drought or flooding, both of which can be harmful to the yield. The introduction of invasive fish species and degradation of water quality are the main ways that mixing pond water and fish with wild fish can have a damaging impact on the environmental sustainability of aquaculture production. Other risks associated with flooding include losing fish from ponds during floods, unwanted species invading ponds, and pond damage from infilling. Furthermore, because fish losses from ponds reduce profits and spread poverty throughout communities, they pose a risk to the social and economic aspects of aquaculture sustainability. It has been noted that increased rainfall may have an impact on the productivity of macroalga such as kelp by introducing different nutrient loadings. Aquaculture productivity may suffer from drought-related water stressors like shortages and deteriorating quality. There will be more disputes over water among the many groups, including aquaculture, agriculture, and businesses, as a result of the anticipated water shortages and competition for resources brought on by climate change. This will have an impact on every aspect of aquaculture sustainability. Further research is necessary to determine how various fish species and life stages, particularly those with significant economic value, will react to modifications in the event of a decreased or changed precipitation pattern.

Aquaculture (Figure 7.3) can also be impacted by storms, waves, and cyclones, mostly in coastal and marine regions(Mohanty et al., 2017). Rough sea conditions can lead to coastal erosion and have an impact on fish and shellfish in cages. The likelihood of aquatic organisms escaping due to equipment failure may also surge due to the increased frequency of storm events predicted for specific seasons in some regions. This could necessitate site relocation or changes in production practices, which could have a serious impact on the social and economic sustainability of aquaculture in these areas. At least 80% of all climate-related disasters are extreme climatic incidences, which have increased recently in a number of regions.

***Figure** 7.3* Aquaculture

Source: https://en.wikipedia.org/wiki/Aquaculture_of_catfish#/media/File:Loading2_U.S._Farm-Raised_Catfish.jpg, *Creative Commons License.*

But it's expected that similar occurrences would occur more frequently in Africa, especially in East and Southern Africa. Storms may enhance nutrient and water column mixing and prevent temperature stratification, which may raise aquaculture productivity. Climate change adaptation refers to a wide range of activities aimed at decreasing the adverse consequences of climate change while utilizing emerging opportunities. It comprises adapting behaviors and policies to anticipated or actual changes in the climate. The process of adapting to climate change can be challenging and filled with issues, especially when conflicting demands for scarce resources lead to frequent disputes. Temperature anomalies, such as shifting air and water temperatures, variations in the sea surface temperature, anomalies in precipitation, rising sea levels, floods, droughts, and cyclones, are examples of physical risks. Researchers identified low oxygen levels, acidification or lower pH values, and salinity variations as significant chemical risks in culture fluids. Pathogens, parasites, eutrophication, and toxic algal blooms are among the biological risks. Furthermore, current scientific knowledge regarding the impact of certain consequences varies and is occasionally limited to cumulative impacts, making adaptation planning nearly difficult in the aquaculture industry.

7.7 STRATEGIES TO LIMIT LOSS

Strategies for modifying the effects of climate change and adapting to it that are both efficient and effective in fostering resilience and adapting to changing climatic conditions may be beneficial for farming communities, ecosystems, and populations in general. The use of gabions to protect vulnerable river banks; the construction of levees/embankments or the use of secondary flood channels and/or retention ponds to minimize or contain the overflow from rivers; the restoration of river bends to manage the hydrodynamics of river flow and the transport of sediment; the installation of groynes to interrupt river flow or revetments to absorb water energy; and the employment of natural flood management strategies can be helpful. Fish grown in cages in ponds, rivers, lakes, or open seas may require regular evaluations of the cage's structure and ability to hold fish in place while preventing loss or escape. All mitigation and countermeasures, however, require capital investment; nevertheless, at the local level, these investments might be out of the reach of many small- to medium-sized producers, and at the national level, they might be too costly for low- to low-middle-income nations, where coordinated strategic plans for regional management would be necessary. Some studies have examined the effects of climate change on aquaculture, categorizing them as either positive or negative consequences, direct or indirect. The majority of these research studies have addressed how aquaculture and fisheries might adapt to climate change in the future, with an emphasis on issues related to yield unpredictability, reduced profitability, and economic uncertainty. Simultaneously, many other studies have concentrated on the impact of climate change on the productivity of fisheries and aquaculture i.e., marine, coastal, and inland aquaculture. The impact of climate change on cage farming and the necessary mitigation and adaptation strategies have not been well studied(Oyebola & Fada, 2020).

7.8 CAGE CULTURE

There aren't many case studies that address how cage culture is affected by climate change and the mitigation strategies that have been released in recent years. The majority of studies originated in Asia, particularly in Thailand, where many losses were experienced by the quickly growing freshwater cage culture. Furthermore, marine cage-cultivation techniques in the temperate zone are hindered by sudden and quick environmental changes brought on by changing weather patterns. Understanding the effects of sudden climate change on cage farming is becoming increasingly important. Research has also attempted to reveal biochemical or molecular changes in cage-cultured species in response to climate and seasonal variations, which had previously been recorded as simple fish deaths. The main reasons why fish die during seasonal temperature variations are changes in feed efficiency, infections that follow heat stress, disease outbreaks in warm water in the tropics, and molecular and biochemical changes in fish. There may also be stratification or declines in dissolved oxygen in the culture environment. In evaluating the impact of climate change on cage farming, it is imperative to take into account the financial damages. Unexpected fish deaths, damage to cages, disease outbreaks, and pollution might cause farmers to have to leave their farming site until it's ready for a new cycle in addition to a one-time financial loss. The loss of natural stocks may also temporarily reduce the availability of seeds/starters for cage farming. Fish stress can be due to various factors such as flooding from high rainfall, which can lead to deformities in cage infrastructure and cage relocation. Additionally, fish stress can be caused by damage and stress from flood debris, collisions with cultured species, and inflow of polluted water and sediments. Conversely, low water flow results in poor water quality because of reduced levels of dissolved oxygen in the water and higher fish densities due to low water levels, which stresses fish and ruins cages. Additionally, a drought may force cages to be consolidated at certain sites in higher densities, which may exacerbate water quality problems and raise the risk of infectious diseases.

7.9 SALMON FARMING

Salmon are cold water fish that cannot survive in warm water. For Atlantic salmon, the ideal water temperature is between 11 and 14°C(Crouse et al., 2022). Within this temperature range, salmon eat well and grow quickly; but, when water temperatures rise beyond 16°C, they get stressed and eat less quickly. If the temperature rises above 23°C, salmon die. Although recorded deaths in farmed stocks occurred at 20°C, it's possible that other mitigating variables also played a role. The greatest risk that climate change is expected to pose to salmon cage farming is sea level rise. Sea level rise, and frequent storms, ocean acidification, and modifications to marine biological processes like the frequency of harmful planktons and increases in parasite populations are other climatic factors that may have an impact on marine aquaculture. Designing risk communications and other risk management actions requires a comprehensive grasp of the concerns that various groups of people have about particular scenarios. In each aquaculture region, it is imperative to ascertain the processes, likelihood, and potential extent of adverse effects on cage culture resulting from climate change events. Using this information, initiatives to lessen the possible effects of climate-related events can be consolidated. For example, the effect of high rainfall and runoff on river cage culture can be explained as an influx of contaminated water or sediments that stress fish, making them more susceptible to disease and raising mortality rates. Significant rainfall and the rains following a dry spell are also important climatic variables; an increase in frequency is a concern related to climate change in this scenario, and riparian bank conditions and watershed land uses are interacting elements. Another illustration of how fast temperature changes affect cage culture in reservoirs is known as thermal destratification. This process exposes fish to low dissolved oxygen stress following abrupt temperature drops, which results in fish mortality. The seasonal shift into the cool or dry season and windy conditions mixing upper and lower water layers are the main climate drivers in this scenario. After learning about the risks associated with climate change and its mechanism, farmers can choose the best management practices to put into practice. Additionally, advantages, operating costs, resource availability (electricity for aerators), and long-term effects on the culture cycle (early harvest/product quality) should all be taken into account when assessing how effective the various solutions are. Some of the main risk management strategies against climate change include aeration, avoiding certain seasons, early harvesting, placing cages closer to banks, lowering stocking density, and transferring fish from cages to ponds. The best economically, socially, and environmentally sustainable management techniques can be chosen after the effectiveness of the risk management options available has been assessed, with the consent of all farmers. An action plan should then be developed, with a focus on large-scale cage farmers or stronger farming communities, as these would be advantageous in managing a variety of risks.

To increase fish production while minimizing environmental impact, climate change adaptation is necessary. While drastically reducing its negative effects on the environment, cage aquaculture productivity needs to increase sustainably. Cage aquaculture adaptability must be adaptable enough for methods to be discontinued, changed, combined, or replaced with new ones. Moreover, adaptive strategies need to fulfill several requirements, such as being robust enough to function well in a range of conditions, flexible enough to adapt to changes in the future, free of undesirable side effects, cost-effective, and supportive of all stakeholders, especially small farmers. Moreover, adaptive strategies need to be effective enough to significantly reduce risk. The public concern over the health of farmed fish, especially those raised in cages, is growing. Although land animals have been domesticated for thousands of years, intentional selection and unintentional domestication in fish are relatively new phenomena. The main focus of selection programs is on characteristics relevant to productivity, including disease resistance or faster development. Fish welfare may face several issues as a result of selection for rapid development. For instance, there might be selection carrying over effects to other inherited qualities that are crucial for welfare because the ability to grow quickly is a polygenically inherited trait. Fast growth in some fish can increase size variance in a cage population, which can negatively impact the welfare/health or productivity of less aggressive fish(Dara et al., 2022). Fast-growing fish are more aggressive. Therefore, although cultured fish are less stressed than animals raised on land due to a lack of domestication, deliberate selection for rapid development may jeopardize the welfare of particular populations. Developing techniques for assessing welfare under cage conditions, identifying farming practices that compromise good welfare, and creating mitigation strategies for such negative consequences are all necessary to guarantee that fish farmed in cages have a good quality of life.

7.10 OPERATIONAL WELFARE INDICATORS

Reliable indicators of the degree to which fish of a particular species and stage have their welfare/developmental needs addressed are necessary when a set of requirements has been determined. The set of indicators known as operational welfare indicators or OWIs, which are very simple and accurate to apply on operational farms, can be divided into groups. Environment-based indicators, also known as input indicators, focus on the surroundings and available resources for fish. These consist of the temperature, salinity, pH and alkalinity, dissolved oxygen and carbon dioxide, total nitrogen, turbidity, and current velocity of the water. They also contain elements of the farming routine, such as the quantity of food provided, the illumination in the cages, and the density of the stocking. To use such OWIs, one must measure the status of each key variable and confirm that it is within the predetermined limits necessary for the survival of the species. Although their nutritional and physical needs have received most of the attention, fish also have behavioral needs. Fish-specific characteristics are used as animal-based indicators. Certain metrics show the condition of the entire group. Body weight can be observed from live fish kept in a cage or from fish that have been slaughtered for sampling. Typically, the sum of these individual ratings is used to evaluate group averages and variability. Interpreting a single result/yield OWI is not always simple. Important qualitative information regarding fish well-being can be obtained from underwater camera footage or behavioral observations made when the fish approach the surface to feed. For instance, some Atlantic salmon attain sexual maturity in production cages despite farming practices intended to postpone breeding, which has a negative impact on their well-being. Although they are frequently challenging and/or time-consuming to gather, quantitative behavioral changes are also informative. In this context, the application of recently developed computer vision systems for precise behavior quantification is growing.

7.11 OTHER FACTORS

Behavior can be a sensitive sign of unsuitable cage situations. Many stressors, like the presence of a predator, can cause a group of fish to lose their regular, polarized spatial arrangement. While general assessment and protection principles apply to all species, extrapolating specific needs and indications from one species to another is obviously unsuitable. All species of cage-farmed fish require the right resources, taking into account the different needs. However, when it comes to food, for instance, there are differences in the nutritional type and optimal presentation of this food. For instance, fish that eat on the bottom, like Atlantic halibut (*Hippoglossus hippoglossus*), (Figure 7.4) grow and feed more effectively when given sinking pellets as opposed to floating pellets(Brown, 2010).

Figure 7.4 Atlantic halibut

Source: https://en.wikipedia.org/wiki/Halibut#/media/File:Fishermen_with_their_halibut.jpg, *Creative Commons License.*

Feed distribution must also be coordinated with natural rhythms, which can differ between and within species, in order to practice good feed management. For instance, the Amazonian species tambaqui (*Colossoma macropomum*), which is extensively farmed in Brazil, exhibits nocturnal feeding behavior, consuming 84.98% of its daily food intake at night, and grows more optimally when fed throughout the night. While a suitable environment is necessary for the good health of all fish species raised in cages, the specifics of that environment are largely determined by the natural environment to which the fish are suited. It is evident that different species, such as freshwater and marine species, as well as tropical and temperate species, have rather different ideal water qualities. Although the specific range of diseases from which they must be shielded varies within these similar parameters.

Many fish grown in cages exhibit a variety of stress management techniques, which might have an impact on how welfare/health is evaluated and safeguarded. For instance, active fish react to perceived danger by actively fleeing, whereas reactive fish react differently. These two types of fish exhibit distinct behavioral indications of panic. *For example*, a distinct pattern of physiological stress response means that low cortisol levels in active fish do not always signify that stress is not there. Stress coping style also affects the kind of health issues that fish raised/grown in cages must deal with. A lot more information is required to guarantee the good welfare of all fish species that are raised in cages, given the wide interspecies variance in welfare needs and health/wellbeing indices. Many people lack access to this kind of information; therefore, much research is required to close this gap and some of it is already in progress. Fish have far more freedom of movement in cages since they are frequently much larger than in tanks and raceways. They are also open constructions, depending on tidal flows and nearby currents to replenish the cage water and remove debris. This, in part, eliminates the need for more technology and makes it simpler to maintain high water quality. In contrast, precise control of ambient water conditions is far more difficult and highly dependent on meteorological factors than tank-based systems. As a result, there are variations in both space and time within a given cage. This could be beneficial from a welfare standpoint within certain limits since it gives fish a choice regarding currents and temperature and constitutes a type of environmental improvement. For instance, during times of high tidal flow, Atlantic salmon in sea cages modify their behavior by moving against the current. But beyond a certain point, it becomes difficult to use input indicators to guarantee fish wellbeing due to less control over environmental factors. Fish kept in cages are easier to observe during routine farming than fish kept in deeper natural ponds. Nevertheless, bad weather will occasionally make it impossible to monitor the results of OWIs because of obstructed access and reduced visibility, for instance as a result of eutrophication and algal blooms. Assessing the welfare of fish raised in cages presents several practical challenges due to the size of the water body that the cages enclose and the large number of fish involved, particularly in intensive cage aquaculture. For instance, in European Atlantic salmon farms, a single cage may be initially stocked with over 200,000 fish(Oppedal et al., 2011).

With an emphasis on safeguarding cage-farmed fish from detrimental organisms, diverse cage designs and equipment have been created to shield Atlantic salmon from the infectious larval stages of different sea species that gather in surface water with good lighting. To physically keep lice and salmon apart, protective cage designs include skirted cages and semi-enclosed containment systems; submerged cages that keep fish from approaching the water surface; and submerged cages with snorkels that let fish access the water surface. Underwater lights can deter salmon from approaching the water surface by utilizing their innate reaction to light. Many fish species visit certain locations on coral reefs known as cleaning stations, where they come across blue-streak cleaner wrasse (*Labroides dimidiatus*), which consume ectoparasites. Because of these behaviors, temperate fish species like lumpfish (*Cyclopterus lumpus*) and corkwing wrasse (*Symphodus melops*), which both consume excreted fish parasites, are farmed alongside Atlantic salmon. Although this method has been shown to be effective in lowering sea lice populations, it has encountered problems, including the dwindling of cleaner populations in the wild, the introduction of diseases, and uncertainties for the well-being of wrasse. Although jellyfish swarms cannot physically harm salmon housed in semi-closed containment systems, their welfare may be indirectly jeopardized if jellyfish clog pump intake. Larger jellyfish can be redirected from fish farm cages by vertical currents produced by plumes of bubbles produced by pumping compressed air through underwater porous tubes; however, this is dependent on depth and local wave/tide regimes.

Using natural behaviors that help wild fish defend themselves against infection and take care of their body surfaces is a unique instance of environmental enrichment that may help prevent infection or diminish disease. Sea trout, for instance, can pick up on the ability to ignore visual cues connected to prior exposure to contagious trematodes. Using cleaner fish to get rid of sea lice is an example of using an organism's natural behavior to lessen its negative consequences. Fish in the wild frequently exploit temperature gradients to regulate their physiological state and enhance the effectiveness of many critical processes, including upstream migration. Common carp elevate their body temperature above normal, which accelerates their growth. Fish that are infected also migrate toward warmer water, exhibiting behavioral fever as opposed to physiological fever. For instance, infected common carp and zebrafish (*Danio rerio*) exhibit behavioral fever, selecting water with a higher temperature(Huntingford et al., 2020). Their elevated body temperature reduces the likelihood of mortality by inducing a potent and targeted immunological response. Research is being done to determine whether behavioral fever, which Nile tilapia exhibits in reaction to infections, may shield them against disease when raised in ponds.

7.12 MAPPING STRESS

Fish that are comparatively untouched by standard farming techniques can be farmed by selective breeding for low-stress response, which has generally proven successful in the common carp and rainbow trout species. One complicated issue here is that selection for decreased stress responsiveness may lead to increased aggressiveness, given the prevalence of stress coping strategies in many farmed animals. To minimize any negative impacts on fish welfare, we should describe detailed regimes for certain farming activities that raise concerns about the wellbeing of farmed fish. These include meticulous planning ahead, ongoing observation and modification throughout the procedures, and suitable post-procedural measures. To lessen the necessity of catching and handling fish for grading and sampling, passive grading has replaced traditional grading. Fish swim freely through an underwater grid, occasionally in response to basic directional cues. In Norway and the UK, commercial grading systems are widely employed in salmon aquaculture.

It is possible to derive behavioral markers of stress from submerged cameras and multibeam sonar, as well as from changes in swimming dynamics. Fish cortisol concentration may be employed, for instance in European seabass, to detect chronic stress. Certain species, such as rainbow trout have skin pigmentation patterns that correspond with plasma cortisol levels and may be utilized as remote indicators of physiological stress and stress coping mechanisms. The use of computer vision and other varied techniques for collecting and analyzing big data sets, along with technological advancements in data collection, are contributing to the increased efficacy of such remote monitoring systems. A large portion of the fish produced to feed the expanding human population would inevitably come from aquaculture due to the depletion of fish supplies. Intensification is probably inevitable in some cage-farming areas for the sake of increased production. To lessen the impact on the environment and prevent conflicts, this may entail the construction of larger offshore sea cages. There will probably be more species diversity in other sections of cage farming, maybe with a focus on omnivorous and herbivorous animals rather than piscivores. Increased research efforts to determine the welfare requirements of these newly domesticated species will be necessary as a result of further diversification. The practice of cage farming will undoubtedly increase the risk of infectious diseases, necessitating the development of new strategies for their management, treatment, and prevention. Veterinarian care in intensive fish farming has long focused on disease management; however, despite advancements in diagnostics and treatments, overall fish farm deaths have not decreased in the last 20 years.

7.13 MOLECULAR TECHNIQUES IN AQUACULTURE

The application of molecular techniques in aquaculture in general is probably going to increase. These will undoubtedly be employed more and more in selective breeding programs. High-throughput genotyping, for instance, has produced genome-wide markers for the primary species raised in cages, making it possible to collect particular traits based on genotypes. Focused selective breeding programs for critical production qualities, such as growth rates, maturation age, stress reactivity, and disease resistance, have already benefited from this(Gjedrem et al., 2012). The tools needed for these kinds of assessments vary depending on the species, but the field is developing quickly, and more species will have access to commercial resources. It has also been shown that genome editing technologies can be used to create genetically superior lines for a variety of farmed fish species; among the qualities that are frequently targeted are greater disease resistance and decreased reproduction. Commercially accessible cage-farmed species include Nile tilapia and Atlantic salmon that have undergone genetic editing to promote rapid growth. Gains from such selective breeding programs may have an adverse effect on fish welfare, as with any farming practices that result in rapid growth. Growing awareness of fish sentience and how it affects fish welfare in culture systems is likely to be one of the research advances motivated by public concern for the welfare of farmed fish. There is probably going to be ongoing pressure to find enrichment strategies that are both practical and cost-effective for use in big fish cages to enhance positive welfare.

Using biosensors that have been created as research instruments to examine fish health on an individual basis is one approach that may be considered(Endo & Wu, 2019). Theoretically, every fish in a cage might be fitted with one of these smart tags to enable precise welfare/health assessment. For cages that are sparingly stocked with high-value fish, like tuna, this may be a viable business option. But since every fish would need to have its tags taken off and be scanned at the point of slaughter, this would also present challenges. Creating cage systems that allow fish to have their unique needs met is arguably the most practical way to protect their welfare across a range of coping methods to stress. This may entail providing access to a variety of temperatures and currents, and a plentiful supply of evenly distributed feed, possibly with the help of the smart feeders. An expanding range of sophisticated sensors are available to aquaculture for the remote monitoring of the aquatic environment, fish welfare, and health. The resulting massive amounts of data can be analyzed using sophisticated tools. These might alter how fish welfare is evaluated and protected, at least in certain industries that rely heavily on intensive cage aquaculture. A growing number of in situ sensors that record factors like temperature, flow velocity, and dissolved oxygen are automatically collecting data on the environment within individual cages. Contemporary cage aquaculture is also utilizing a growing amount of publicly available data regarding several pertinent environmental variables. To extract and analyze data relevant to fish welfare in this scenario, a variety of analytical and modeling approaches are needed due to the size of the data sets from these sources. Digital technology can provide early warning of hazardous conditions outside of cages.

Chapter 8

CLIMATE CHANGE AND INSECTS

Contents

8.1 INTRODUCTION

Among the most attractive fossils are insect corpses preserved in amber to the exact size of a hair, which have captured the public interest as a hypothetical source of dinosaur DNA. In lacustrines, ponds, lagoons, and marine environments, insects are frequently preserved. Body mass, size, and shape, in addition to environmental factors like oxygen content influence how well-preserved a species is. The rich and varied fossil record of Coleoptera serves as an example of the taxa that are most likely to have medium body sizes and small wings. Changes in fossil insect abundances are likely more related to environmental changes than they are to actual changes in insect population sizes. Therefore, taxonomic, morphologic, evolutionary, and biogeographic investigations have been the main areas of concentration for palaeoentomological study. Gathering reported taxon occurrences and entering them into the Paleobiology Database has taken a significant amount of work and time. Our comprehension of the diversification and extinction of insect families throughout geologic times is greatly aided by these data. Based on statistical analysis, it can be observed that the record of insect fossils is inconsistent in both space and time, with a high concentration of occurrences spanning from the Carboniferous to the recent. Only the amber deposits in Tanzania and the Dominican Republic maintain low-latitude Cenozoic insect faunas; the majority of these sites are in the Northern Hemisphere and are located at mid- to high latitudes.

Currently, plants account for over 80% of all biomass on Earth. Even though leaves only make up 7% of plant biomass, their biomass still exceeds that of all animal biomass(Antar et al., 2021). Trees frequently lose and regrow their leaves, which produces a large number of possible fossils. Therefore, it should come as no surprise that fossil leaves cover a larger area and span a longer time period than fossil insects do. They also capture important time intervals and areas devoid of fossilized insects. Each leaf fossil usually represents tens to thousands of years. Leaf fossils are typically maintained in river or lacustrine habitats, which makes comparisons among fossil floras easier. Paleoecology, plant community composition, and climate may all be reconstructed using the same leaf fossils that were used to study herbivory. In addition to providing direct evidence of prehistoric food webs, insect herbivore damage on leaves provides information on the insects that are present in an ecosystem. Herbivory damage is easily separated from physical damage and detritivory change. Cells near the injured location grow or multiply if leaf tissue is removed while the leaf is still living. Compared to detritivores or physical processes, insect herbivores usually target specific plant tissues, organs, or species and produce more regular patterns and morphologies of damage.

8.2 DAMAGE TYPES (DTS)-THE CONCEPT

Damage patterns unique to individual insects, such as cusp markings or pupal chambers, can be noticed. The key to insect (and other) damage types on compressed plant fossils is used to classify insect damage into morphologically distinct, diagnosable damage types (DTs). The functional feeding groups (FFGs) that DTs belong to are piercing, sucking, hole feeding, margin feeding, and surface feeding. DTs can be further classified as specialized DTs, which are usually produced by insects that eat a few closely related plant species, and generalized DTs, which are typically produced by polyphagous insects(Currano et al., 2021). When it comes to dating insect species, physically unique DTs, such as leaf miners and galls, can sometimes be linked to an insect family or even the genus.

This is typically not feasible, though, as certain DTs can be produced by a variety of unrelated insect species. The fact that certain insects are capable of producing several DTs further complicates things. There may be a positive association between the number of insect species found feeding on a plant host and the amount of DTs produced by these insects.

Therefore, DT diversity serves as a substitute for insect herbivore diversity and can be studied at the level of individual plant hosts or flora. Unbiased insect damage censuses are used to assess herbivory. Each leaf from a fossil site that can be identified and is more than 50% complete is scored for the presence or absence of each DT. To record regional variation in plant composition and herbivory, it is advisable to excavate many sites, if feasible: a total of 1,000 leaves should be dug. The metrics frequency i.e., percentage of leaves with herbivory damage, diversity (number of DTs), composition, and intensity (% of leaf area damaged) are frequently used to quantify herbivory damage. Metrics related to herbivory can be studied on both individual plant hosts and floras.

Total damage, specialized damage, and mining and galling damage are among the common metrics that are reported. Sixty-one fossil insect damage surveys on floras dominated by Cenozoic angiosperms have been carried out as of 2021. The best-sampled time periods are the Paleocene and Eocene, and two-thirds of all sites are in mid-latitude regions of either Eurasia or North America. The availability of paleofloras that may be investigated for herbivory is not reflected in the time and spatial dispersion of censuses of fossil insect damage. Instead, it is based on the geographical regions and areas of interest for the few paleontologists who have conducted these assessments. It is recommended that palaeobotanists include insect herbivory in their repertoire of palaeoecological investigations of ancient flora.

8.3 ISOTOPES

Analysis of oxygen isotopes in deep-sea fossil foraminifera that live at the bottom provides the best record of global climate change during the Cenozoic(Veizer & Prokoph, 2015). The oxygen isotopic composition (δ18O) of their carbonate shells is determined by the seawater δ18O and the temperature at which they formed. The volume of polar ice has an effect on seawater δ18O because ice sheets preferentially incorporate $_{16}$O, hence an increase in ice volume increases δ18O. Therefore, the δ18O record indicates fluctuating ocean bottom-water temperatures during the Paleocene and early Eocene, before massive continental ice sheets formed, with the amount of variations correlating to changes in the surface temperature. The middle to late Eocene saw the development of massive Antarctic ice sheets, and the δ18O record shows a combination of changes in temperature and ice volume. Paleoceanographers have created and assembled more records since, improving temporal resolution and revealing both short- and long-term climate changes. Some researchers made important contributions to our comprehension of the dynamics of the global climate by introducing the framework of trends and aberrations in climate. The main force behind the slow changes in climate is plate tectonics. The collection of deep-sea foraminiferal δ18O has a temporal resolution of one sample every 2200 years for 0–34 Ma and one sample every 4400 years for 34–67 Ma(Nagasundaram et al., 2014). Glacial-interglacial cycles during the Pleistocene and Holocene were controlled by orbits, and plant and insect migrations have been extensively studied, especially after the last glacial maximum. Rapid disruptions to the global carbon cycle are climate aberrations.

8.4 FOSSILS

In many instances, historical temperature and precipitation may be estimated using the same fossil leaves that were examined for insect damage. Studies on contemporary forest ecosystems have shown relationships between morphological traits of woody dicot leaves and climate factors. These correlations can be utilized to infer ancient/historical climates from fossil leaf collections. For instance, average leaf size and mean annual precipitation (MAP) are correlated, and the percentage of species with smooth vs serrate margins is correlated with mean annual temperature (MAT) in subtropical and temperate forests. A number of techniques have been put forth to use the climatic tolerances of closest living species to estimate historical climate parameters like MAT, MAP, growing season temperature and precipitation, growing season length, and mean cold month temperature. Taxonomic composition can also be used to reconstruct climate for more recent floras (Neogene). Tools for reconstructing historical climates include geochemical indicators such as pedogenic carbonate nodules, enamel, speleothems, leaf waxes, and microbiological biomarkers. Some researchers separated the Cenozoic into distinct climatic states, building on earlier paleoclimate research. Greenhouse conditions prevailed over the first 10 million years of the Cenozoic (66–56 Ma), with temperatures > 5 °C higher than they are now. By the beginning of the Eocene (56 Ma), Earth had transitioned into a hot-house state, with temperatures > 10 °C warmer than today and sudden global warming events (aberrations) or hyperthermals, the largest of which being the PETM. Around 60 Ma, temperatures started to climb continuously. The Cenozoic era's highest sustained temperatures were recorded during the EECO (Early Eocene Climatic Optimum). Over the course of the remaining Eocene, temperatures progressively dropped, first returning Earth to the greenhouse between 47 and 34 Ma and then moving to the Cool-house close to the boundary between the Eocene and Oligocene (34 Ma). The significant decrease in δ18O during the early Oligocene was probably brought on by both substantial Antarctic ice sheet growth and cooling. After the Miocene Climatic Optimum (17–14 Ma), δ18O steadily dropped, indicating a mix of ice sheet extension and cooling. During 3.3 Ma (the middle Pliocene), the present-day Ice-house state was formed, with fully developed ice sheets at both poles.

During the warm, humid Paleocene and early Eocene, most of the earth was covered with closed-canopy, broad-leaved forests, as shown by fossil plants and geologic deposits like bauxites, evaporites, and coals that provide climate information. Megathermal rain forests significantly contracted during the late Eocene and early Oligocene cooling, especially in the northern hemisphere(Fernández et al., 2021). Most of the continent above 45°N, South America, and the ice-free regions of Antarctica were covered in cold temperate woods around 30 million years ago. The distributions of moist forests were further constrained over the Oligocene and Miocene by the ongoing cooling and aridification of the continental interior, while the Miocene Climatic Optimum probably permitted a brief expansion. During this period, there was an increase in grasslands, which further fragmented forests and provided insects with new habitats and food supplies. Understanding the relationships between insect dynamics, flora, and climate change is becoming more and more dependent on molecular phylogenies, which are dated using both insect fossils and analytical insect damage.

8.5 PHYLOGENETIC ANALYSIS

According to phylogenetic studies, plant abundances and diversities were impacted by Cenozoic temperature changes, which in turn had an impact on the diversification and extinction of related herbivore taxa. Diffuse cospeciation is characterized by the dispersion of plant geographic ranges caused by climate change, the diversification of insect herbivores in tandem with their plant hosts, and the colonization of the newly separated relatives of the host when geographic ranges reunite. Insect speciation events are caused by increased abundance and dispersion of the host plant, which is generally driven by climate changes, under resource abundance-dependent diversification patterns. There are other ways to investigate the connections between diversification events and climate change, including the integration and expansion of insect fossil records, paleovegetation reconstructions and models, and paleoclimate data and models. Because they are the only tools available to examine extinction events, fossil studies also supplement molecular phylogenetic studies. Long-term variations in temperature and vegetation may have an impact on insect diversity globally. Due to more consistent and unbiased sampling and established insect families, analyses of fossil insect diversity are typically conducted at the family level as opposed to the genus or species level. Species richness peaked in the early Cretaceous, at about 125 Ma, fell through the Late Cretaceous, then regained in the Early Paleogene. It is believed that the diversity of families now is either less than or comparable to that of the Early Cretaceous. According to phylogeny-based diversification analyses and insect diversity curves, during the warm Paleocene–Eocene, there were relatively high origination rates and net diversification at the family level. These rates then began to decline near the Eocene–Oligocene edge, as Earth transitioned from greenhouse to Coolhouse conditions.

8.6 RECORDS

Although the data are highly irregularly distributed in time and space, no global trends in insect damage diversity or frequency have been identified across the entire Cenozoic. The fossil record makes it abundantly evident that over thousands to millions of years, temperature variations have an impact on herbivory in natural ecosystems. Reconstructed MAT and total, specialized, and galling damage diversity have statistically significant relationships in the Cenozoic global herbivory database. Correlations were also found in an older dataset that included Palaeozoic through Cenozoic sites. The Cenozoic database spans 64 million years, and the locations are distributed from the Arctic to Antarctica. Regional datasets spanning from mid- to high latitudes show much stronger connections. Reconstructed MAT and damage diversity indicators show strong, positive associations when compared to nine sites in the Bighorn Basin, Wyoming, USA, spanning 6 million years and including the PETM and the EECO. Reconstructed MAT and damage frequency showed weaker but still substantial positive association/correlations. Comparable outcomes were discovered at a series of Neogene sites in Iceland, where damage diversity decreased as a result of prolonged cooling. Only two Paleocene and nine Eocene sites have data on leaf area damage, and the data for these sites shows a weak positive correlation with temperature. Increased herbivory at high temperatures may be caused by a number of mechanisms. High damage diversity during periods of global warming may be a sign of either migration or speciation events(Bush, 2002). Damage diversity is a substitution for the diversity of insect herbivores. Due to the Bighorn Basin's mid-latitude location and the evidence of plant taxa migrating northward during the Eocene as temperatures warmed, it has been hypothesized that the northward migration of a variety of thermophilic insect herbivore species is responsible for the elevated damage diversity within the basin during warmer intervals. The global herbivory database is dominated by mid-latitude locations as well. With warming, insect species may have relocated northward or higher in elevation, while with cooling, they may have gone southward. Factors like insect speciation and extinction need to be taken into account when analyzing damage diversity patterns. Broad-leaved, closed-canopy forests reached their maximum geographic spread during the warm Early Paleogene, and mountain ranges may have provided stronger biogeographic barriers than they do in colder temperatures. It's probable that a combination of the species-area effect and allopatric speciation causes an increase in global insect herbivore diversity, and consequently impacts diversity, during Warmhouse and Hothouse intervals. It is possible that during cooling periods, mesothermal forests shrunk, leading to insect herbivore extinctions, which are represented in the fossil record as reductions in damage variety.

8.7 CLIMATE EFFECTS

The direct influences of climate change on insect herbivores as well as the indirect effects through changes in plant nutritional content are what cause increased damage frequency and intensity with warming. As long as warming doesn't go beyond a species' thermal tolerance, it accelerates the metabolism and growth of insects, shortening their development times and reducing their death rates. This allows for more generations to occur annually. Each of these elements can raise population densities, which in turn raises the frequency and intensity of herbivory. Plants cultivated in experiments with higher CO_2 frequently have lower C:N ratios, which may have an adverse effect on insect population densities but may also result in compensatory feeding. The Bighorn Basin paleofloras indicate that high temperature and CO_2 favored legumes and other plants that frequently have symbioses with N_2-fixing bacteria. The fossil record, however, shows large fluctuations in plant species composition with climatic change. Plants that possess N_2-fixing symbionts, such as legumes, have higher leaf nitrogen levels than non-N_2-fixing symbiont-containing plants. Adding high-nitrogen plant matter to soils probably gives rise to larger and more varied insect populations by acting as a source of nitrogen that is biologically accessible. The fossil record shows significant positive correlations between the frequency, intensity, and diversity of both total and specialized damage, as well as between the prevalence of N_2-fixing plants and these factors. The fossil record shows that increased herbivory is the ultimate consequence of warming and elevated CO_2 over extended periods of time in natural systems. The glacial-interglacial cycles that have defined the last 2.6 million years, or the Quaternary, are caused by periodic variations in the Earth's orbital parameters(Winckler et al., 2004). Over many decades, temperature changes of up to 7 °C resulted in the expansion and retreat of the northern hemispheric ice sheets, which covered a major portion of the continental landmasses in the mid-latitudes. Researchers have been able to map the spatial distributions of individual plant species, plant associations, and biomes at millennial-scale resolution for the past 21,000 years due to extensive research on vegetation change in the Northern Hemisphere during the most recent deglaciation and the compilation of hundreds of fossil pollen records. There have been significant changes in the biogeographic ranges of plant taxa, individualistic responses of those taxa to climatic change, and the development of communities i.e., combinations of taxa that are not currently observed.

Fossils of Quaternary insects are found in anoxic, wet environments such as peat bogs and lake bottom sediments. Because of how well preserved their original chitin exoskeletons are, a thorough comparison with extant faunas is possible. The majority of Quaternary fossil specimens correspond to current species, which may come as a surprise considering the abrupt and drastic changes in flora and climate that were previously recorded. Important research on fossilized beetle fossils from glacial and interglacial deposits shows that species can change their geographic ranges to follow temperatures that are suited for them. Today, species from warmer, late-glacial sites in Britain are found in continental Europe, Turkey, and Northern Africa; taxa from British glacial deposits are found in northern Eurasia, northern North America, or high-elevation regions of Asia. Similar evidence of elevational and latitudinal range alterations linked to climatic change has been found in New Zealand, where fossil beetle collections dating back 1.3 million years have been used. As populations moved across complicated terrains, they would have periodically split up and reassembled, creating mixed gene pools and stabilizing selection. The mobility of the species that did become extinct was restricted by either photoperiod requirements or topography. The idea that long-term temperature patterns are more likely to be strong creative forces in macroevolution than glacial-interglacial cycles is supported by the Quaternary insect fossil record. The majority of Quaternary insect fossils are found in mid- to high-latitude regions of the north, and ecological and evolutionary trends may be different at lower latitudes. DNA analyses show Quaternary insect speciation episodes in regions of Europe that were not affected by glaciers, indicating that diversification events may have also happened at lower latitudes. Some researchers compared the damage caused by insect herbivores on Quercus Oak leaves from 3 Ma to present woods in southwestern China(Castagneyrol et al., 2019). They found that there was a 50% increase in DTs in the modern/latest forests, and only one gall DT was extinct. Subtropical and tropical fossil faunas and floras, as well as the identification of new locations, should be investigated. Integrating the highly comprehensive palaeobotanical record with the Quaternary insect fossil records is another interesting area of investigation. The differences in ecological and evolutionary trends between generalist and specialist herbivores, predators, scavengers, and detritivores may be the subject of future studies.

8.8 PHENOTYPIC PLASTICITY

Importantly, phenotypic plasticity can play an important role in the survival of species because it enables populations to respond to changing environmental conditions within very short timescales. For instance, it has been shown that in certain insect species, such as *Drosophila* spp., plastic responses that alter the onset and length of diapause promote population persistence. Similar to how ontogenetic flexibility in features like hardness and temperature acclimation might be important for an insect's persistence, these physiological responses are necessary for survival in the face of abrupt temperature fluctuations. Genomic techniques are beginning to identify the molecular genes that control how significant traits in insects are expressed in response to climate change. The best example comes from the extensive work on *D. melanogaster* and close relatives, where a strong body of empirical data now allows assessment of the genomic basis of thermal plasticity and its repeatability. Furthermore, thermal plasticity in gene expression levels, including the number and types of genes that are variably expressed between temperatures has been shown by transcriptome studies conducted on insects. Once more, *D. melanogaster* has been the subject of extensive research on the impact of temperature on transcription, and most studies report that this species exhibits a wide variety of flexible responses to temperature. The great adaptability of *D. melanogaster* and many other Drosophilidae species is likely what has allowed for the different shifts caused by climatic change in this group. There is little overlap between the potential loci mediating variation in plasticity, suggesting that the genetic basis of thermal plasticity is heterogeneous and occasionally unique.

Figure 8.1 *Bicyclus anynana*

Source: https://en.wikipedia.org/wiki/Bicyclus_anynana#/media/File:Squinting_bush_brown_(Bicyclus_anynana_anynana).jpg, *Creative Commons License.*

8.9 GENETIC STUDIES

Studies of the thermal transcriptional system in the African savannah butterfly *Bicyclus anynana* (Figure 8.1) revealed pervasive gene expression differences only between seasonal phenotypes, and the almost complete absence of intra-population genetic variation for plasticity within a season, in contrast to the extensive genetic variation present in *Drosophila* spp. This suggests that in the event of a future mismatch between seasonal phenotypes and climate, a lack of genetic variation within populations could have disastrous consequences. In *Ischnura elegans*, the genes that were elevated in response to cold stress varied in core and edge populations, which is in line with the quick response to colder conditions(Lancaster et al., 2016). The use of integrated genomic high-throughput techniques on subjects exposed to varying heat regimes has demonstrated that changes in the epigenetic regulation of gene expression cause a multitude of phenotypically flexible responses. According to research on the cold-tolerant goldenrod gall fly (*Eurosta solidaginis*) and the cold-tolerant goldenrod gall moth, genes involved in freeze tolerance such as DNA methyltransferases (dnmt), histone acetyltransferase, and demethylation enzymes are differentially expressed, suggesting epigenetic regulation. Similarly, it was discovered that in the invasive white fly *Bemisia tabaci*, chromatin remodeling and differential expression of dnmt1 and dnmt3 were critical for heat shock survival. The best candidate genes that cause changes in thermally sensitive phenotypes are usually those that have both expression patterns and functions related to temperature; these genes may be identified in the future. The global distribution of insects is changing as a result of climate change. In fact, research conducted in the Northern and Southern hemispheres has shown that various insect groups have undergone significant changes.

8.10 GWAS FINDINGS

Expansion results in new species interactions and community compositions. One such interaction is a higher rate of hybridization between previously allopatric species. Given that more and more insect species are being known to exhibit some degree of niche conservation, it is critical to comprehend how adaptable certain species are to climate change. To enhance the modeling of evolutionary processes that incorporate measurements of gene flow, population dispersal, and genomic load, all essential for forecasting the fate of species across landscapes, researchers have resorted to landscape genomic analyses as part of this effort. Furthermore, methods like genome-environment association (GEA) studies make it possible to connect certain genomic variants to novel environmental traits and offer proof of environmental adaptation. According to research on damselflies, dispersal evolution is a significant part of the process, and some species may be pre-adapted to range shifts by exhibiting high dispersal propensities already at their range core. This is because GEA studies are used to map genomic variants to environmental conditions at range margins. Moreover, the utilization of genome-wide association studies (GWAS) has been implemented to identify the genomic areas accountable for range shifts related to climate change. A GWAS investigation conducted on populations of D. melanogaster revealed twelve SNP alterations linked to the critical thermal maximum (CT_{max}). It's interesting to note that these studies also showed that the minor allele in the majority of these SNPs raised the upper temperature limit, indicating the existence of raw genetic variation for heat tolerance. Additionally, research on range shifts has shown that populations are probably responding to changes in their environment at varying speeds depending on where they are in their range. At the edge of their range, where populations are more likely to be isolated, species are typically expected to show declines in genetic diversity but increases in genetic differentiation. Indeed, several Odonata species have been discovered to share this common trait, a loss of genetic diversity and/or an increase in genetic divergence in populations close to range boundaries. However, historical and continuing range shift processes mitigate the intensity of the species-specific patterns.

8.11 VARIATIONS

Variations in the local selection regime influencing populations are a manifestation of alterations in the surrounding environment. Changes in selection regimes like these will eventually lead to genetic alterations and, finally, adaptability. Local adaptation is thought to be the most significant scenario to ensure the long-term existence of species in a warming climate(Valladares et al., 2014). It happens when animals have greater average fitness in their local environment compared to individuals from elsewhere. However, in situations when fitness experiments are unavailable, care must be used when interpreting genetic population differentiation in important traits as adaptive. It is not possible to establish a direct relationship between genetic alterations and fitness in the absence of strong evidence, necessitating further research to measure the consequences of phenotypic fitness.

There is a wealth of evidence supporting genetic changes in key traits over time or throughout a species range. These changes include phenology tolerances to high temperatures brought on by climate change and increased metabolic energies associated with surviving on a warmer planet. Using a combination of single-locus and a polygenic multilocus method, the scientists also discovered a genetic signature of adaptation to the recently encountered thermal regimes, both at a single SNP and in a group of covarying SNPs. These days, whole-genome techniques are being used to investigate how insects are responding to climate change, especially small-genome insect species. Hundreds of SNPs linked to two distinct and thermally induced diapause symptoms were found by analyzing thousands of *D. melanogaster* and subjecting them to various thermal tests/trials followed by whole genome sequencing. These investigations have also demonstrated that, in addition to SNPs, large inversion polymorphisms also contribute to temperature responses and exhibit predictable frequency fluctuations over seasonals. This finding has been consistently observed in Drosophila spp. across continents. Similar results were also found in *Coelopa frigida*, where diverse natural and sexual selection patterns on males and females lead to balancing selection and polymorphism maintenance across heterogeneous landscapes. These studies also found that a large inversion polymorphism was linked to multiple environmental variables, including temperature.

8.12 COPING WITH GLOBAL WARMING

Global warming is a major driver of insect extinction today as it has been in the past. For survivors to react quickly, they must first develop a physiological system that has been in place for millions of years that allows insects to withstand heat. Such equipment requires sophisticated heat perception and temperature interpretation by the nervous system, antioxidants, a well-coordinated heat coma, an efficient but expensive stress response, and a host of cellular, metabolic, and thermoregulatory reactions. In addition to being a regulator of heat acclimation, an irreversible increase in heat tolerance acquired during an individual's development and hardening, a reversible acquired tolerance, or adaptation, these physiological mechanisms also give insects their heat tolerance. Research on insects has determined heat tolerance, or the ability to withstand high temperatures, by measuring the upper lethal temperature (ULT) or the critical thermal maximum (CT_{max}), which is the highest temperature at which an insect can survive without losing its ability to control its muscles. It can also be quantified as the amount of time an insect can remain normal at a specific high temperature. This reaction is an outcome of a confluence of physiological processes that enable the insect to recognize an extremely high temperature and react accordingly. Insects rely on evaporative cooling and external (cuticular) coloration in addition to internal physiological defense mechanisms to withstand heat(Bodlah et al., 2023).

These mechanisms, when paired with specific behavioral reactions, enable insects to tolerate high temperatures. In addition to varying with age, sex, body size, and condition, heat tolerance in insects is also sensitive to environmental factors like contaminants, photoperiod, symbionts, parasites, and even the environment of their parents. The global temperate zones are most affected by climate change in terms of ambient temperature. The biophysical principle of living at basal higher temperatures, which states that even small temperature increases require disproportionate increases in metabolism, potentially increasing energy expenditure and oxidative stress, is the main reason why insect diversity is more threatened in the tropics. This is partly because more biodiversity exists at these latitudes. Nevertheless, despite the potentially large range of mechanisms underlying their heat reactions, tropical insect species are still little understood. As ectothermic and poikilothermic species, insects depend on outside heat sources and have variable body temperatures. Insect body temperature, however, seldom coincides with the surrounding air temperature due to the ability of insects to thermoregulate through behavioral and physiological mechanisms.

8.13 THERMOREGULATION IN INSECTS

Insects that are subjected to heat thermoregulate based on their behavior. Examples of such insects are *Colias eurytheme* butterflies and *Enallagma doubledayi* damselflies, (Figure 8.2) which constantly orient to minimize solar radiation exposure and reduce energy loss at high temperatures.

***Figure 8.2** Enallagma doubledayi*

Source: https://en.wikipedia.org/wiki/Enallagma_doubledayi#/media/File:Enallagma_doubledayi_16234044.jpg, *Creative Commons License.*

When the temperature outside is high, insects can also thermoregulate by flapping their appendages to move cool air about. Because melanin lowers cuticular permeability, dark phenotypes heat up and cool down more quickly than lighter phenotypes, and they may be more resistant to dehydration. Insects can physiologically regulate their body temperature by modifying the amount of melanin in their cuticula during development. Insects at low latitudes or altitudes are typically lighter than species from cooler locales as an adaptive response.

In Diptera, Hemiptera, Orthoptera, and Hymenoptera, evaporative cooling is a process in insects that aids in lowering body temperature(González-Tokman et al., 2020). For instance, at exceptionally high ambient temperatures, honeybees from the Sonoran Desert vomit droplets of honey that cool the body. Even though famine and dehydration do not appear to affect the ability of *Drosophila melanogaster* to tolerate heat, small insects are not predicted to thermoregulate by evaporative cooling due to the little amount of water they can retain. Using the tracheal and circulatory systems, insects can control the transfer of heat from colder to warmer body portions. Heat tolerance necessitates a coordinated response of physiological and neural defense mechanisms, such as heat shock proteins, biogenic amines, and neuroendocrine factors, which collectively conform to the stress response. These mechanisms must first involve heat sensation, processing, and interpretation of thermal information. An integrated response to heat also includes hormonal changes, thermoregulation, antioxidants, and metabolic alterations. Sensation of the right temperature is the initial step in the reaction to heat. This is accomplished via peripheral nervous system neurons that are linked to the central nervous system, which organizes a suitable response.

8.14 METABOLISM

Insect metabolism increases under excessive heat to the point that it becomes impossible for them to find cooler homes and control their tracheal movements, which can lead to death. In response, insects go into a state known as heat coma, which is a reversible response including brain and muscular shutdown that increases the likelihood of survival by preventing cellular damage and energy depletion. The ventilatory central pattern generator (VCPG), a neural circuit in the central nervous system, is found in insects that have evolved to withstand extremely high temperatures, such as the desert locust *Locusta migratoria*. This circuit is housed in the metathoracic ganglia, uses the tracheal system to transport and remove O_2 and CO_2 and to facilitate evaporation, which cools insects. The neurotransmitter serotonin, which rises in heated insects and appears to play a crucial role in thermoprotection during insect heat coma, protects the VCPG before heat stress.

Insect metabolism rises with temperature before entering a heat coma. Increased metabolism becomes extremely expensive for insects exposed to heat because it requires increased ion pumping and protein turnover, which results in energetic depletion and death. Resting metabolic rate, on the other hand, represents 50% of total energy expenditure. Insects from warm regions, like the tropics, are more vulnerable because of the nonlinear relationship between temperature and metabolic rate. This is because their baseline temperatures are greater and even slight temperature rises necessitate a corresponding rise in metabolism. Aerobic metabolism, or respiration, increases with temperature, but ventilation and circulation may not be able to keep up with the need for oxygen, which could lead to mortality. According to the oxygen and capacity-limited thermal tolerance (OCLTT) hypothesis, animals die at high temperatures due to a lack of oxygen, which is more restricted in water than in air because oxygen in water drops sharply with temperature.

Temperature-related increases in metabolism may lead to an increase in the formation of highly reactive free radicals such as hydrogen peroxide (H_2O_2), hydroxyl radical (HO), and peroxide anion (O_2^{2-}), which can cause damage to cells. Oxidative damage to biomolecules results in cell death when antioxidants are not available. Certain antioxidants are derived from the diet, whereas others are created internally. Insect antioxidant defenses against heat include the following enzymes: glutathione (GSH), glutathione reductase (GR), lactate dehydrogenase (LDH), transferrin, ascorbic acid, proline, and superoxide dismutase (SOD), peroxidases, catalases (CAT), and glutathione-S-transferase (GST). The homogenates of heat-stressed Lepidoptera, Diptera, Coleoptera, and Hemiptera, as well as some specific organs like the testes of heat-exposed silkworms *Antheraea mylitta*, have higher levels of these antioxidants and total antioxidant capacity. Furthermore, oxidation resistance genes are upregulated in the entire body of heat-stressed Asian citrus psyllids (*Diaphorina citri*)(Mafa et al., 2020). Other antioxidants that are gained from diet include carotenoids, ommochromes (from tryptophan), and melanins. However, it is unknown how these antioxidants relate to heat resistance. Climate change may require dietary modifications in addition to endogenously produced antioxidants since it may alter insect diets and, consequently, the availability of antioxidants for certain insects. There is currently less data regarding insect heat responses, however complementary processes such as uncoupling proteins in the mitochondrial membrane may assist minimize the generation of reactive oxygen species or ROS and oxidative stress when exposed to heat. As heat-acclimated larvae induced more differentially expressed antioxidant genes in response to heat shock than the unacclimated larvae, acclimation and adaptation to heat are also linked to antioxidant defenses in Lepidoptera. However, more experimental research is required to determine the role of antioxidant responses in insect plastic and adaptive responses to heat. Heat exposure can be fatal because it denatures proteins. All living organisms, from bacteria to vertebrates, can withstand extreme temperatures and other forms of stress due to chaperone proteins called heat shock proteins (HSP) that have a range of molecular weights. HSP works by attaching itself to other proteins to prevent denaturing at temperatures that would otherwise be fatal.

The neurological system, the stomach, and the fat are among the tissues from which HSP protects cells in hot insects. Most HSPs are elevated in response to heat, however, in certain areas, such as the testis, ovary, and fat body of *Bombyx mori* (Lepidoptera), they are downregulated (HSP70 expression is decreased). HSP overexpression is followed by a decrease in its levels since the heat shock response is an energy-demanding process. On the other hand, the fly *Drosophila arizonae* and the desert locust *Locusta migratoria* have high levels of HSP70 and do not respond to heat upregulating this HSP in locations with exceptionally high temperatures, such as arid and semi-arid regions of Africa, Asia, and America. Hormones and biogenic amines (BA) play a role in the neuroendocrine stress response. Following heat stress, the insect fat body releases a variety of brain chemicals, including octopamine, dopamine, serotonin, adrenaline, and. BA functions as neurotransmitters and neuromodulators in neural cells and can also be released into bodily fluids to operate as neurohormones. BA is frequently released in response to additional stressors such as cold, disease, and malnutrition. Biogenic amines control gene transcription and the $Na^+ K^+$ ATPase pump, which is essential for responding to various environmental signals and transmitting brain impulses(Armstrong & Meldrum Robertson, 2006). Many temperature-sensitive enzymes, including alkaline phosphatase and dopa-decarboxylase, are required for the release of BA. However, it is still unknown how climate change will affect these enzymatic processes in insects. Similar to how glucocorticoids function in vertebrate stress responses, neuroendocrine factors like Adipokinetic hormone (AKH) release BA and cause insects to release stored energy. Heat causes AKH receptors in honeybees to diminish, which indicates a reduction in carbohydrate mobilization in stressed bees. However, more study is required to comprehend the function of neuroendocrine systems in insect responses to heat.

Although it is essential for insect survival in the heat, the endocrine stress response during heat tolerance is likewise less understood. Insulin and gonadotropin hormones are the involved hormones. The insulin/insulin-like growth factor signaling or IIS pathway controls the synthesis of antioxidant enzymes like SOD during insect diapause and provides tolerance to cold, oxidative stress, and pathogenic infection. This pathway is mediated by transcription factors like Forkhead box O (FOXO), which is also present in vertebrates. FOXO also controls insulin signaling and energy metabolism in Drosophila in response to heat stress. Juvenile hormone (JH) and ecdysone, two gonadotropin hormones that enhance insects' endocrine stress response to heat, are likewise synthesized via insulin signaling. Since gonadotropin hormones and insulin signaling have only recently been linked to insect heat tolerance, non-model insects should also be studied in terms of their neuroendocrine response to heat. Microbial endosymbionts, some obligatory and some facultative carry out or support some insect tasks, including heat tolerance. However, these symbioses are threatened by climate change, which could lower host performance at high temperatures. In *Nezara viridula*, an insect exposed to simulated climate change, for instance, gut symbioses fail, resulting in severe fitness problems at temperatures 2.5 °C above the normal(Oliver & Higashi, 2021). The lack of a heat shock gene in the obligate endosymbiont bacteria Buchnera decreases heat tolerance in the host, which suggests that changes in bacterial heat sensitivity alter heat tolerance in the insect host. Aphids are arguably the most studied system on this topic. Additionally, the facultative symbionts *Serratia symbiotica* in the pea aphid shield Buchnera from heat stress in addition to the host. Facultative mutualisms can also exist. Microbes can be population-specific since the microbial communities within insect hosts are shaped by both the local environmental influences and the evolutionary history. In most insects, the role of the microbiome in heat tolerance is still unknown, but it may be more significant than previously thought.

Many physiological systems related to heat tolerance are revealed by transcriptome analysis, however most of them have not been well investigated. For instance, transcriptome research in *Apis mellifera* showed that, in comparison to a control temperature of 25 °C, exposure to 45 °C results in the overexpression of 334 genes and the downregulation of 100 genes(Gao et al., 2020). When compared to a control temperature of 25 °C, exposure to 40°C causes the overexpression of 1,275 genes and the downregulation of 1,222 genes in *Glyphodes pyloalis*. When compared to 25 °C, 476 genes in the parasitoid wasp *Tamarixia radiata* are activated and 26 genes are downregulated in response to 38 °C. In addition to HSP and antioxidants, these genes also contain detoxification and vitamin digesting pathways, repressors of fat and carbohydrate metabolism, and even immune response genes like the Toll-like receptor signaling pathway. The physiological response appears to differ throughout the examined insects in terms of gene regulation; nevertheless, transcriptome analyses of the heat shock response are still missing for the majority of insect orders. More research should be done on these mechanisms. To determine the gene expression modules implicated in heat tolerance, appropriate experimental designs and gene co-expression studies are required. The actin cytoskeleton of heat-acclimated crickets, *Gryllus pennsylvanicus*, has changed as a result of other methods, such as fluorescence microscopy, but the function of the cytoskeleton in insect heat responses is still unknown. Insects that lose body lipid reserves in response to heat exposure clearly demonstrate this.

8.15 HEAT EFFECTS

High temperatures in this era of global climate change invariably interact with other environmental stresses including pollutants, food scarcity, and new parasites and infections. When these stressors are combined, their interactions can either be synergistic or additive, meaning that the total effect of the stressors is what matters. It is generally anticipated that when global temperatures rise, the severity of stressor effects will increase. For instance, heat waves only cause sublethal damage to damselflies (*Coenagrion puella*), resulting in immunosuppression and metabolic depression. However, only when combined with a lack of food do the effects of heat prove fatal for larvae exposed to the agricultural pesticide chlorpyrifos, indicating the critical role that nutrition plays in the reaction to heat. HSP70 and the enzyme acetylcholinesterase, which is essential to synapsis, do not appear to be engaged in the response to the synergism, whereas the cellular electron transport system appears to regulate the synergism between coupled stresses. Exposure to heat may appear advantageous when managing other stressors. For example, dung beetles or *Euoniticellus intermedius* exhibit greater heat tolerance when artificially parasitized, and mosquitoes *Culex pipiens* exhibit reduced sensitivity to chlorpyrifos after prior heat exposure. It is unclear from the available research whether this outcome stems from cross-tolerance to both stressors or heat. The persistence of a species in warm environments will be mostly determined by the synergistic effects of heat with other stresses besides life-cycle balance with heat tolerance. The physiological mechanisms that underlie the survival of individuals from natural environments warrant specific examination. In these populations, which evolve with multiple stressors, the mechanisms responsible for responding to heat alone may differ significantly from those responsible for responding to a combination of stressors. Insects are a diverse group of animals and have played a vital role in terrestrial ecosystems since their emergence some 480 million years ago. This example of successful evolution demonstrates their adaptability and capacity to alter with their surroundings. However, insect populations are facing important challenges due to the rapid changes in surroundings during the last 150 years or so. These findings are concerning since shifts in insect abundance have immediate effects on ecosystem services and other aspects. Consequently, entomologists, evolutionary biologists, and ecologists all are very interested in the mechanisms that enable insects to adapt to changing environments(Kellermann & van Heerwaarden, 2019).

Insects generally have some ways to adapt to changing climates. Tracking changes in environmental conditions through changes in geographic distribution ranges is an option, if it is feasible. These changes are not the result of deliberate migration, but rather of certain populations suffering extinction as a result of increasingly unsuitable environmental conditions, while other populations disperse and establish in previously unsuitable locations. Although alterations in distribution caused by climate change have long been known about and are frequently seen, most insects are unlikely to be able to relocate because of habitat fragmentation and man-made or natural obstacles to dispersal. Insects may rely on phenotypic flexibility when relocation is not an option. Throughout their lives, insects have developed defense mechanisms against environmental change, and they can modify several facets of their physiology, morphology, or behavior to adapt to changing climate conditions. Such within-generation reactions have the potential to protect populations against abrupt environmental change if they are adaptable. Through adaptation, insects may also tolerate abrupt changes in their environment. Insects can undergo genetic changes in response to unfamiliar environments. Under climate change, a wide range of biotic and abiotic factors can dictate or impact insect phenological shifts. These factors can be broadly classified into organismal drivers and environmental drivers. On the one hand, phenological shifts may result from modifications to environmental signals that affect insect phenology, such as temperature, precipitation, photoperiod, etc. Even though the responses of a species to temperature change may be similar, they may differ in phenological shifts because of the varying warming rates across latitudinal and elevational gradients if the species experiences different climates at different latitudes and altitudes. Organismal factors, or variations in the sensitivity of a species or population to environmental change, can also cause phenological mismatches or changes. Even if they suffer the same rate of warming, species or populations that are more susceptible to temperature changes should exhibit larger phenological fluctuations than those that are less sensitive(Forrest, 2016).

8.16 PHENOLOGY

The phenological synchronization between herbivorous insects and their host plants, as well as between pest insects and their natural enemies, might be affected by variations in temperature sensitivity across sympatric species belonging to various trophic levels. Insects are ectothermic, meaning they are unable to control their internal body temperature. In response to changes in external temperature, insects' development and seasonal timing, or phenology, exhibit plasticity. The development and metabolic rates are both temperature-sensitive, following the thermal performance curves (TPCs). As a result, depending on whether and to what degree the warming exceeds the optimal performance temperature, these rates may exhibit either positive or negative reactions to temperature increases. Warmer weather often has the potential to quicken insect development and provide more advanced spring phenology. However, certain insect species need to be exposed to extremely cold temperatures for a prolonged periodto end their diapause, which suggests that warming would also likely lengthen the diapause and cause delayed spring phenology. Warming, however, might cause changes in the phenological shifts because of the wide ranges in phenological flexibility between species and taxa. The degree to which phenological shifts are impacted by climate warming may depend on the plasticity of a population or species in response to temperature change, or thermal sensitivity. More sensitive populations or species would show more pronounced phenological shifts. For instance, the temperature sensitivity pattern demonstrates that phenological shifts are larger or occur more quickly in animals that are more sensitive to temperature fluctuations. Species with a higher annual generation count tend to have shorter life cycles and will therefore be more vulnerable to global warming than species with a lower annual generation count. Butterflies that have a higher number of host plants demonstrate a higher degree of phenological advancement compared to those that have fewer host plants. Significantly, the population abundance and demographic responses to climate change are probably going to depend on how flexible phenology is.

To maintain synchrony with their resources, for instance, insect species or populationswithgreaterphenologicalflexibilityinresponsetovaryinginterannual climates may be able to thrive or be less negatively impacted in abundance under recent climate change. Positive demographic responses of multivoltine

Lepidoptera species to climate change may result from phenological flexibility, such as flexibility in voltinism and active period. Many insect species possess the ability to adapt their phenological responses to swift changes in climate through evolution. For instance, to adjust to the longer growing season brought on by climate change, the mosquito *Wyeomyia smithii* exhibits delayed diapause induction, which is genetically regulated by photoperiodic response(Bradshaw & Holzapfel, 2017). To keep up with the temperature variations across their geographic ranges, the UK butterflies with lower dispersion capacities advanced their spring phenologies, indicating substantial local adaptations. Furthermore, there are many instances where populations within species exhibit phenological modifications, such as the timing of diapause induction and the annual number of generations along climatic gradients. These examples demonstrate the ability of species to adapt widely to a variety of climate conditions. Yet, limited gene flow and genetic divergence between populations are typically expected to be linked to local adaptation. This suggests that genomic alterations to future environments differ significantly, which is likely to result in local adaptation under the strong directional selection of rapidly warming climates.

8.17 DORMANCY AND OTHER EVADING TECHNIQUES

Certain insect species have developed evading techniques to adjust to fluctuations in interannual temperatures or growing/dormant season durations. If a previous phenological window is missed, these tactics may allow a subset of the populations to track synchrony with their feeding/mating resources. If the species are unable to achieve the requirements for their diapause/dormancy termination, they can choose to adapt to the specific climatic conditions in different ways. Insufficient chilling brought on by warmer winters may result in prolonged dormancy or delayed diapause termination. On the other hand, prolonged cooling brought on by colder winters may result in another dormancy cycle, which allows adults to emerge a year later. It is frequently discovered that extended dormancy comes with fitness expenses, such as a higher risk of mortality and lower adult fertility, suggesting some limitations on the evolvability of evading techniques.

Insect diapause/dormancy is often triggered by photoperiod, while additional signals like temperature, precipitation or drought, and the quality of the host plant can also influence this process. Therefore, although the photoperiod will remain steady at a given latitude, climate change may affect the commencement and/or termination of insect diapause and consequently lead to advanced or delayed phenology by altering the temperature and/or precipitation/drought.

Novel environmental conditions will therefore result in selection pressures and quick plastic or evolutionary diapausing reactions in insect phenology. Warmer autumn temperatures generally cause autumn phenology, like diapause induction, to be delayed. Conversely, warmer winter or spring temperatures can cause diapause termination to occur earlier and/or accelerate post-diapause development, which in turn causes spring phenology to advance. But in temperate regions, winter warming can also cause spring emergence to be delayed since there is not enough to end diapause. On the one hand, spring phenologies, such as adult emergence and spring migration, have advanced in many insect species as a result of climate change. For instance, from 1978 to 2015, the French aphid spring migration accelerated by over one month. The degrees of phenological shifts, however, vary based on the organismal drivers as well as the environmental drivers. Not all insect species exhibit accelerated spring phenology; some exhibit delayed spring emergence as a result of insufficient winter chilling for the termination of diapause. Consequently, distinct or even opposing variations in phenologies may arise from seasonal warming(Primack et al., 2009). In this instance, the phenological results for certain species can be influenced by the timing of the temperature shift as well as the degree of the temperature increase. For instance, in the univoltine butterfly *Anthocharis cardamines*, warming over the winter may produce delayed spring phenology because the pupae need cumulative winter chilling to break obligate pupal diapause, and greater winter temperatures postpone the commencement of diapause termination. Increases in annual generations can result from climate change that advances spring phenology while delays fall phenology, thus extending insect growing seasons. In both functionally univoltine and multivoltine species, the extended duration of adult insect activity may result in more generations annually. For instance, it was shown that, in Central Europe, the frequency of second and subsequent generations of multivoltine moth and butterfly species had significantly increased since 1980, with 44 out of 263 species exhibiting an increase in the number of generations per year. *Dendroctonus ponderosae*, the functionally univoltine beetle, is not diapause-prone. Warmer spring and summer temperatures can therefore cause a summer generation in Colorado by extending their activity seasons, accelerating their temperature-dependent development, and advancing their flight season by more than one month. Even so, a lot of species continue to be obligately monogamous in spite of rising temperatures.

Increases in voltinism brought on by phenological changes may have intricate population-level demographic effects. Climate change will not affect the photoperiod; nevertheless, it can alter the rates of development that are reliant on temperature. This could strengthen or weaken the relationship between crucial photoperiods for diapause induction and diapause-sensitive developmental phases. Warmer temperatures may lead to an increase in the abundance of multivoltine insects if they can successfully produce new generations each year that can enter diapause at a specific developmental stage and withstand winters. Conversely, warming may result in inefficient reproduction and excessive energy consumption before entering diapause, which could cause high mortality and a decline in population abundance or even lead to extinction when the additional generations are unable to complete their development during the extended growing season. For instance, a generation of wall brown butterflies, *Lasiommata megera*, may result from increased autumn temperatures(Klop et al., 2015). These species may not be able to reach the instar, when they can enter the age-specific larval diapause, due to the substantial mortality associated with such fall warming. Insect population-level consequences, however, will vary based on biological responses to environmental signals, as well as spatial and temporal factors.

8.18 OTHER ASPECTS OF HEAT STRESS

Heat stress may slow down the development of ectotherms throughout the growing season, however not all trophic web species will experience a growth decline in this process. Temperature rises can either speed up or slow down development. Different species within the same communities may experience varying degrees of diapause during unfavorable seasons; some species may continue to have a significant percentage of diapausing members while others may become fully active. Furthermore, the majority of insect species contain facultative or obligatory bacterial symbionts, whose functions can be changed or modified in response to temperature increases. This can have negative effects on insect survival or enhance the resistance of insects to temperature stress through either the indirect effect of temperature on the symbiont itself or direct symbiont-mediated host protection. Phenological mismatches between distinct trophic web species may arise from these consequences. Higher trophic levels that rely on the ability of lower trophic levels to adjust to these changes are expected to be more affected by climate change. Phenotypic variations have differing effects on interspecific interactions. In mutualistic relationships, mismatch will usually have detrimental effects on both species. From an evolutionary perspective, trophic mismatch is predicted to persist for brief periods in some situations, as the species will either adapt or vanish after a few generations. The short duration of phenological mismatches and the need for long-term ecological data are two aspects that may account for the difficulties in detecting mismatches caused by climate change. To be able to identify the true phenological disparities caused by climate change, it is specifically necessary to have a pre-climate change baseline, which is an assessment of the level of synchrony that a particular interaction had before recent climate change. The match-mismatch hypothesis (MMH) offers a conceptual framework for explaining how phenological variations affect population dynamics in antagonist relationships. The growth rate reaches its maximum when its phenology is almost identical to that of its resource, and it diminishes as the mismatch rises. This model also demonstrates how the degree of the phenological mismatch with their resources limits the viability of the species. However, a perfect synchronization between two species may lead to overexploitation of the resource, which in turn may cause both species to decline over time. Shorter phenologies are less susceptible to climate change. Phenological mismatches have different effects on multivoltine and univoltine organisms. A substantial component of consumptive interspecific interactions is voltinism and changes to voltinism may have an impact on the total amount of phenological overlap(Knell & Thackeray, 2016).

Furthermore, some studies have suggested that univoltine species may be phenologically mismatched and undergo continuous selection in an attempt to adjust to shifting seasonality; in contrast, multivoltine species may be less affected and undergo reduced selection, at least in the later generations of the season. In insects, mutualist interactions and interference are examples of non-consumptive interspecific interactions. The degree of interspecific temporal overlap, which is a good indicator of interaction intensity in mutualist relationships, determines the risk of phenological divergence after climate change. A mismatch is more likely to happen when mutualism takes place during seasonal, facultative, and generalized interactions. This condition is also more likely to occur at higher latitudes and altitudes when there is a large seasonal component. On the other hand, natural selection has created interacting organisms that are mutually obligatory and specialized to respond to abiotic stimuli similarly and hence should be more resilient to phenological mismatch. Generalist species are less likely to be selected to react to signals that are similar to their resources, but if they can compensate for the loss of some interactions through other interactions, the mismatches may not have as much of an effect on their fitness. In fact, there are instances when the level of specialization is unequal. For instance, a specialized plant that flowers sooner may experience a mismatch with its pollinator, which could result in a failure of pollination. However, if the pollinator can make up for the mismatch with other floral resources, the mismatch may not affect it.

8.19 PHENOLOGICAL MISMATCH

Climate change can affect plant-herbivore interactions in two ways: either it can cause a phenological mismatch that protects the plant from herbivores, or it can synchronize the plant with its herbivore and cause herbivore outbursts.

Plants that mismatch in their phenology can avoid damage from herbivores. The environment can also cause herbivore outbreaks by synchronizing the herbivore with its host plant. The timing of budburst is crucial for young larvae in the system that includes the spruce budworm (*Choristoneura fumiferana*) and its host plant, the white spruce (*Picea glauca*), because the suitability of leaves rapidly diminishes after budburst(Lirette & Despland, 2021). The tree defoliated as a defensive tactic, which causes an earlier budburst (defoliation) induced mismatch with herbivores. This phenological mismatch has shielded the plant up until this point, but winter warming decreases the mismatch and the synchrony raises the likelihood of outbreaks.

Butterflies and moths often improved their phenologies about plant–pollinator interactions. But butterflies do not progress their phenology at the same rate as their flowers, unlike bees. Mismatches have not often resulted in total decoupling, most likely due to natural selection constraining plant–pollinator interactions from synchronizing. In insects, phenological mismatches can also lead to evolutionary changes. Winter warming causes caterpillars to hatch before foliage becomes accessible, which results in a decline in the population of the winter moth (*Operophtera brumata*).

This selection pressure caused the hatching time of the caterpillar to be delayed, decreasing the asynchrony and demonstrating that certain species may adapt to climatic change. Few phenologically mismatched conditions cause mutualistic interactions to decouple. A morphological mismatch could be more likely as a result of global warming. A specific plant species' association with its pollinator may be disrupted.

While the impacts of mismatches on seed production are frequently used to assess the fitness consequences for plants, pollinator fitness is still not well known. Phenological mismatches between pollinators and plants, however, can seriously harm pollinator populations and fitness. Males and females may be influenced differently at the species level. For instance, in the Osmia genus, warmer temperatures caused the males to react differently than the females, causing the males to develop during the flowering period and the females to develop later.

This could result in fewer mating and fertilized eggs being produced. Furthermore, because they depend on a single plant species or family to feed their offspring, specialized pollinators are typically more vulnerable than generalist pollinators. Because two species in strong relationships also interact with other species, and these contacts can either be reinforced or eliminated, a phenological mismatch between two species may have an impact on many trophic networks in a particular area.

Figure 8.3 *Parthenolecanium quercifex*

Source: https://en.wikipedia.org/wiki/Parthenolecanium_quercifex#/media/File:Oak_Lecanium_Scale_imported_from_iNaturalist_photo_30467972_on_16_March_2024.jpg, *Creative Commons License.*

Parthenolecanium quercifex(Figure 8.3) species were twelve times more common on oak trees in the warmest urban locations than in the cooler ones(Dawadi & Sadof, 2023). There is a phenological mismatch since the major parasitoid species does not advance egg laying at the same rate as the scale insect, which oviposits earlier in warmer climates. This does not change the percentage of scale insects that are parasitized, but the number of eggs produced by parasitized insects from warmer climates was double that of individuals from colder climates, and the number of eggs produced by non-parasitized individuals did not alter between the environments. Studies show that whereas urban warming causes a pest to develop earlier, the phenology of the parasitoid population that feeds on it does not progress. This mismatch is linked to higher egg production, which most likely increases the number of pests. Season length has a significant role in determining resilience at the community level, with groups that have shorter interaction seasons potentially being the most vulnerable. Mismatches in the interactions between plants and phytophagous insects or pollinators might result in the formation of new biological networks and communities. Certain groups will lose current relationships, but other communities may find new ones. In the cereal aphid-parasitoid community of Western France, part of the aphid species *Metopolophium dirhodum* population is now reproducing parthenogenetically in cereals during the winter, and part of the parasitoids *Aphidius ervi* and *Aphidius avenae* are no longer diapausing. Nutrient cycling may also be impacted by phenological mismatches between interacting species. The increase in caterpillar populations at the start of the season may lead to increased oak herbivory, which in turn causes a decrease in above-ground leaf biomass and a weakening of these trees' capacity to store carbon and nitrogen (N). This would happen if the bird migration does not coincide with the peak of caterpillar biomass. In the case of mutualistic interactions, if apple trees are not pollinated, the carbon they would otherwise devote to fruit production could instead be used for growth and storage, increasing the plant's capacity as a sink of carbon. An increasing number of studies have established the existence of mutualism breakdowns associated with climate changes at the evolutionary time scale. Since all mutualisms are a part of broader ecological interactions, an evolutionary mutualism might break down due to the loss or gain of a single species, which can be brought about by phenological mismatch. Phenological mismatches can occasionally, even in antagonistic relationships, have no negative effects on the community for some reasons.

Chapter 9

BIRDS AND CLIMATE CHANGE

Contents

9.1 INTRODUCTION

Many simulations indicate that the distribution of certain diseases will vary as a result of climate change. Depending on a range of variables and the peculiarities of each disease, experts looking into the impact of climate change on bird-parasite interactions continue to anticipate an expansion of the geographical range of many diseases and a possible decrease in others. Furthermore, it is still anticipated that parasites may achieve increased transmission over longer periods of time if the global climate gets warmer and wet, with increases in virulence likely to follow. It is important to emphasize that both hosts and parasites are living organisms that can alter and adapt, which makes it challenging to forecast exactly how they will react to changes in their surroundings. For example, a large portion of the life cycle of many parasites is spent apart from their vertebrate hosts; consequently, the environment has a substantial impact on the parasites during this time. Under certain conditions, the growth and survival of a parasite and, consequently, its transfer to a vertebrate host, may be impacted by climate change. The hosts' physiology, behavior, and population size may alter as a result of climate, making it simpler or harder for parasites to finish their life cycles. Parasites can also move hosts in reaction to a decrease in the number of their original hosts and an increase in the number of alternative hosts. A unique scenario is presented by introduced host species. They can typically outcompete endemic host species because they are frequently parasite-free or resistant to endemic parasites.

9.2 CLIMATE EFFECTS

The varying influence of various factors, such as environmental constraints, host resistance, interactions between parasites and their predators, or interactions between several of these variables, means that we cannot, in any case, expect to observe a linear relationship between environmental change and parasite populations. Both hosts and parasites may experience various events throughout their life, including interactions with one another, that are impacted by climate change. For instance, rainfall and temperature have the potential to directly or indirectly affect the breeding season by affecting the availability of food. Variations in the timing of the food supply and the breeding season can have significant effects on the health of the parasite or host. Climate change may cause warmer winters or early springs, which could shift the timing and amount of peak parasite transmission and have an impact on host health. Therefore, it is anticipated that climate change will have an impact on the phenology of hosts and parasites, ultimately changing how they interact. Cuckoos (*Cuculus canorus*) and their hosts are an example of how phenological shifts in response to climate change can impact bird–parasite interactions(Yun et al., 2020). Because they will have fewer opportunities for successful parasitism, cuckoo species will eventually go extinct.

Cross-sectional investigations, which evaluate associations between phenological mismatch between parasite and host at the population level and climate conditions, showed that cuckoos could adapt their breeding phenology to match that of their magpie hosts. By contrast, low climatic influence and high individual consistency were found in the phenology of magpie hosts in longitudinal analyses that looked at within-individual variation in breeding phenology. These results imply that the host's adaptation to climatic conditions is not the cause of the phenological mismatch between cuckoos and magpies at the population level.

9.3 TRANSMISSION DYNAMICS

There aren't more instances that show how alterations in bird parasite phenology are related to climate change. Nonetheless, several investigations into several tick species that can serve as hosts for birds have demonstrated or predicted phenological variations linked to environmental factors. For instance, in northeastern North America, *Ixodes scapularis* is the primary tick vector of *Borrelia burgdorferi*, the agent responsible for Lyme borreliosis, as well as other tick-borne zoonoses(Halsey et al., 2018). While humans and small mammals are the primary hosts of this parasite, new studies have revealed that birds can serve as reservoirs for the parasite and aid in the spread of tick-borne diseases. The temperature has an impact on these ectoparasite seasonal phenology. As a result, the degree of seasonal synchronization among the various tick stages may be impacted by variations in climate. Moreover, the diseases that tick larvae spread may vary depending on their seasonal changes. The transmission of infections will be aided by asynchrony between nymphs and larvae because infected nymphs can infect a population of susceptible hosts that can then pass the pathogen on to larvae and complete the transmission cycle. Nonetheless, synchronization between nymphs and larvae facilitates the spread of viruses that cause transient diseases in their hosts.

Shorter-lived, less efficiently transmissible, and infections became more virulent under some conditions of seasonal tick activity, despite the fact that long-lived, highly transmissible, and low virulence pathogens were active. Through the impacts of temperature on tick seasonality, some researchers concluded that climate change may have an impact on the evolutionary processes of infections carried by the tick. Scientists examined the seasonal synchronization of *I. scapularis* nymph and larval stages using data collected over 19 years. In comparison to years when these months were colder, the phenology of nymphal and larval ticks advanced by over three weeks in years when May and August were warmer. More synchronization, nevertheless, was not noticed. Warmer temperatures (October) were linked to fewer larvae feeding alongside nymphs in the spring of the following year. They postulated that the timing of nymphal and larval activity will advance by 8–11 and 10–14 days, respectively, on average based on these findings and expected warming by 2050. Climate change should therefore either sustain or promote the spread of persistent diseases while suppressing the growth of bacteria that do not cause chronic infections. Certain parasite stages can develop quickly in warmer environments, which increases their frequency and intensity in host organisms. For example, the first intermediate host of trematodes, snails, produces more cercarial substances when the temperature rises, and the trematodes develop from these hosts more quickly. Therefore, trematode cercarial output may increase due to global warming, which could affect hosts, especially bird hosts.

Higher temperatures also shorten the cercariae life span. In addition to directly damaging birds through schistosomiasis infections, which are linked to alterations in climate, these parasites can indirectly damage birds by decreasing the populations of amphipods on which they feed. The main reason these trematode parasites have been researched is that they can cause cercarial dermatitis, sometimes known as swimmers' itch, in people who swim in lakes. This disease is not communicable but does induce a skin rash. Because bird host populations are spending more time in lakes and altering their migratory patterns as a result, the rising frequency of this disease is linked to climate change. The prevalence and intensity of these parasites have increased in lakes that had previously never or seldom produced the disease due to the increased presence of birds and the rapid generation of cercariae at warmer temperatures. Trichobilharzia, Bilharziella, and Giantobilharzia are known to parasitize birds and are the cause of cercarial dermatitis in Europe. In their avian hosts, these parasites also induce serious diseases. Another such example that could help us understand the rise in parasite infection incidence linked to temperature increases is the *Heterakis gallinarum*. This parasite affects chickens and other bird species. It spreads by eggs that have been previously deposited in the environment and developed into larvae. Temperature increases speed up the growth of larvae, which may lead to a rise in the frequency of this parasite infection. Heterakis should be closely watched when the climate shifts since it could have an impact on the poultry industry(Collins et al., 2022).

9.4 IMPACT ON AVIAN INFECTIONS

Many investigations have established a connection between environmental factors and avian parasitic infections. Humidity seemed to be significant for trypanosomes, but temperature was the most significant predictor of Plasmodium prevalence. The likelihood of malaria infection was higher in Berthelot's pipits (*Anthus berthelotii*) (Figure 9.1) living in the Canary Islands, with higher minimum temperatures. According to the mathematical models, seasonal temperature variations have an impact on the establishment of the insect vector (blackflies) that spreads the malaria-like blood parasite Leucocytozoon, in the Colorado Mountains. It is anticipated that rising temperatures will have an effect on the incidence and distribution of insect vectors, extending their latitudinal and altitudinal ranges and, ultimately, influencing the prevalence of the blood parasites they carry. The prevalence of Plasmodium has risen in tandem with rising global temperatures, particularly during the past 20 years. In a bird community in the Australian Tropics along an elevation gradient, the prevalence of various blood parasites has also been positively correlated with temperature.

Figure 9.1 *Anthus berthelotii*

Source: https://en.wikipedia.org/wiki/Berthelot%27s_pipit#/media/File:Anthus_berthelotii_-Gran_Canaria,_Canary_Islands,_Spain-8.jpg, *Creative Commons License.*

Both the rise in vector abundance with temperature and the faster development of blood parasites within vectors at higher temperatures have been proposed. The scientists estimated that for every 1°C increase in temperature, there would be a 10% increase in parasite prevalence. In New Zealand, a comparable association between blood parasites and altitudinal gradient has also been documented. Ectoparasites, on the other hand, maybe an exception to this rule as their life cycles depend more on the surrounding temperature for development and completion. As a result, they might be more directly impacted by climate change than endoparasites, which are less impacted by weather once they are inside their avian host. Since certain ectoparasites have close relationships with their hosts, it makes sense that the direct consequences of climate change would have minimal influence on them. However, some research demonstrates how ectoparasites are impacted by climate even in those situations. Mite intensities on their hosts may be impacted by the effects of climate change along altitudinal gradient. Many researchers have connected climatic events to the incidence and prevalence of the parasite diptera *Philornis downsi*. Despite feeding on fruit, adult flies lay their eggs in bird nests, where the resultant larvae consume the tissues and blood of the young flies. This results in significant death rates in certain bird species. In a year with heavy rainfall, researchers discovered higher concentrations of infection by this parasite in nests, but fewer parasites in a year with low rainfall. But rather than a direct result of climate change on the parasite, they concluded that this effect is more likely to be caused by an increase in clutch size during years with increased rainfall.

The mean *Philornis torquans* intensity in Argentinian birds was positively correlated with high average maximum temperature and increased rainfall(Antoniazzi et al., 2011). Both the prevalence and number of parasites were shown to have increased over time, along with an increase in temperature. However, there was no clear relationship between these characteristics. The findings indicated that the apparent effect of temperature on parasites was mediated through an effect on hosts since temperature can significantly affect host characteristics including body condition and clutch size, which are also connected to parasite infection.

A rise in temperature does not always translate into an increase in the frequency and quantity of parasites. Instead, this depends on how temperature relates to other climatic factors like rainfall and the kind of parasite and its unique life cycle requirements. Furthermore, parasites may benefit more from rising host temperatures than from any direct climatic consequences. Range expansion is one of the key hypotheses that relate host-parasite interactions to climate change. Terrestrial vertebrates are increasing their ranges two to three times faster. Even though diseases and parasites normally follow hosts as they expand, there is frequently a lag in timing since they are not present in host invasion areas before expansion. Birds are more mobile than other vertebrates and may therefore easily expand their range. Furthermore, following climate change, migratory species and those introduced by humans are more likely to grow quickly. Studies have looked at the relationship between the expansion of avian host ranges and climate change, either directly or indirectly. For instance, changes in the distribution of avian malaria are generally anticipated to occur both latitudinally and altitudinally. In Alaska, avian malaria is already effectively spread, and it is anticipated that as temperatures rise due to climate change, transmission will also expand. This will facilitate the emergence of new parasites and vectors in those areas.

9.5 IMPACT ON MIGRATION

Climate change is predicted to cause several viruses that afflict migratory birds to expand their geographic ranges; however, the processes behind these effects are currently unclear, and additional ecological research is required. Several factors can influence hosts during range expansion, and more study should be done. Host-parasite interactions are influenced by virulence, which is defined as the net effect of parasites on their hosts(Poulin & Forbes, 2012). A parasite's virulence can vary depending on the host. Therefore, when new host-parasite interactions arise during range expansions, climatic change may have an impact on virulence. The virulence of parasites infecting recently arriving hosts in the extended areas may rise due to climate change, or novel parasites may infect native hosts in these areas. If warmer temperatures enable parasites to finish their life cycles more quickly, hence boosting parasite density, then an increase in virulence is anticipated in both scenarios. Similar mechanisms might similarly impact introduced species and their parasites. There are, however, very few instances of elevated pathogenicity in birds caused by climate change. Variations in physical condition and clutch and brood sizes were connected to variations in parasitism and temperature. Scientists propose that these parasites are influencing their host fertility and physical state, maybe through indirect temperature impacts. West Nile virus (WNV) has also been shown to exhibit increased virulence in migratory bird hosts, where new, extremely virulent strains have emerged as a result of range extension. As in the case of the WN02 strain, which originated in North America and is more effectively spread by mosquitoes, the virulence of these strains may be correlated with climate. Under laboratory conditions, the strain's transmission differential increases with temperature, which makes sense given that it multiplies more quickly than other strains. In certain instances, reduced virulence is noted as temperature rises and parasite populations or activity decreases. For instance, ticks use some strategies to locate their hosts, and they are less active in this pursuit at higher temperatures.

Birds may adapt as a result of changes in the interactions between them and parasites brought on by climate change. It is well-recognized that variations in body temperature can impact the way the immune system reacts to infections. However, the immunosuppressive effect of stress is eliminated when species are previously acclimated to temperature conditions. In any case, there hasn't been much research on the connection between thermal stress and immunosuppression that causes a disease episode or increases susceptibility, despite the fact that many studies have connected the start of epidemic outbreaks to abrupt changes in temperature. Some studies examined the impact of temperature variations along an altitudinal gradient on blood parasite infections and major histocompatibility complex (MHC) variation. Researchers concluded that the association between immunity and temperature is indirect and mediated by parasite abundance, although parasite intensity was related to both temperature and MHC variability. Climate change may have an impact on the level of parasite pressure, which determines the effectiveness of the protective behaviors of birds against parasites. For instance, when ectoparasite infection in the pied flycatcher (*Ficedula hypoleuca*) nests is manually reduced, female grooming and nest cleaning efforts decrease. Therefore, the amount of time birds spend practicing hygiene will probably be impacted by the rise in parasite abundance brought on by climate change. Nonetheless, certain diseases can readily adjust to new environments, resulting in fresh host-parasite dynamics that facilitate the spread of diseases. The West Nile virus has infected new mosquito species in North America and adapted to bird species there. Other possible instances of parasites adjusting to new conditions are migrating European cuckoos that adjust to species that are accessible for nest parasitism, or the impact of Philornis parasitism on birds in the Galápagos Islands(Bataille et al., 2018). The process by which organisms form new associations with other species as a result of the traits that they carry at the time they encounter the new condition is known as ecological fitting, and it should be considered in the case of generalist parasites introduced into new areas/hosts that are favored by climate change.

9.6 PARASITE DYNAMICS

Moreover, the ability of parasites to adjust to varying temperatures could potentially mean the difference between their success and failure in finding a new host. Birds catch our attention because they are easy to observe and are visually appealing. They are the subject of extensive research for some of these other reasons. Avian malaria is another disease that is receiving a lot of attention, most likely because of its similarities to human malaria. The primary motivation for studying even fewer well-studied parasites is their possible direct consequences on humans, like in the instance of cercarial infections. Further research is necessary to determine how climate change affects the distribution, frequency, severity, virulence, and coevolutionary interactions of avian parasites. Because birds and parasites are significant ecological components, these findings will also be essential for people. The majority of research uses field data to create models, including ones that forecast how future temperature changes can affect bird-parasite relationships. Experimental methods that generate empirical data to enhance these models are still lacking, though. It is also possible to investigate bird-parasite interactions in a variety of climate scenarios, which could help us better understand how these relationships change in response to climate change. Our understanding of bird-parasite relationships has undoubtedly increased with the addition of other variables in addition to climatic variables. This has demonstrated that even closely related parasites can respond differently to similar environmental conditions. In addition to host migration and landscape characteristics, the prevalence and richness of parasites were primarily correlated with climate. Furthermore, even when infecting the same host species, scientists discovered that different variables had distinct effects on each parasite type.

9.7 INVADING SPECIES

The possible effects of parasitism and climate on the survival of invading species have not received much research. However, in the event of climate change, the spread of non-native species would be impacted, particularly if these species managed to evade common parasites or if climate change might even serve to facilitate the entry of infections by invading species. It's also hard to predict how avian hosts could react to new diseases. Future studies may concentrate on whether birds can adjust to the new problems brought about by the introduction and adaptability of new infections as a result of climate change. The ability of some bird species to adjust to the introduction of new infections will probably be crucial to their survival. Many genes, species, and ecosystems are affected by climate change. Up to 1.2°C has already been added to the global temperature since preindustrial times. According to projections, there would be a rise in disease transmission and extinction of species as a result of global warming. Birds have advanced their breeding or migration phenology. All trophic levels are expected to be impacted by climate change, but the responses of communities and ecosystems to it have only got the attention. It is not reasonable to assume that multispecies interaction networks, such as those involving predation, parasitism, and pollination, have linear interactions. Many other species engage in interactions with them, and the nature of these interactions can be either positive or negative. Because interaction strength includes the flow of energy through an ecosystem, which is crucial for community stability in particular, it is also functionally significant in relation to ecosystem processes. It is anticipated that climate change will also have an impact on the size of species interactions.

Species will be impacted by climate change through changes in how ecosystems function like pollination, mutualism, and parasitism than by the change itself(Schweiger et al., 2010). Compensatory mechanisms found in complex networks of biotic interactions have been shown to mitigate the effects of climate change on species, but their complexity may also exacerbate those effects. Therefore, understanding and projecting the effects of climate change on the distribution of organisms may not be possible by analyzing a single variable for a single life phase of a few species. Given that species frequently exhibit their own response to temperature, it is reasonable to expect that climate change will influence the temporal and geographical relationship between interacting species at different trophic levels. Range changes may also have a domino effect on how ecosystems work and how communities are organized. Therefore, it is likely that the effects of species range shifts will be more significant than those that result from simply adding or removing species from ecosystems. These alterations could result in ecological disasters such as ecosystem phase shifts. It is especially difficult to forecast how such changes would affect ecosystems, which offers a challenge for ecologists. It is therefore becoming increasingly evident that species-level and community-level responses might be expected to differ from one another, and that it is rare to be able to predict the response of one species from that of another, especially if the latter is associated with a different trophic level. It is imperative that we employ a mechanistic approach that takes into account species interactions in order to comprehend and forecast the responses of species to environmental variability. Thus, one of the main concerns of both scientists and managers is understanding how climate change will affect the organization of ecosystems and communities. Climate change may have a substantial impact on population regulation at the population level, which begs the question of how biotic and abiotic factors differ in their roles in influencing ecosystem structure and population regulation. The documented impacts of climate change on species distributions have given rise to the theory that biotic factors may play an important role than abiotic factors.

The impacts of climate change on interactions between hosts and parasites or between plants and herbivores seem to be varied. In the absence of alternative hosts, climate change frequently disturbs the synchrony of host-parasite phenologies and distributions, phenologically isolating emerging parasitoids from host eggs and possibly leading to localized extinctions that could free host species from parasitism. Because predator-prey interactions are a major cause of animal mortality and are therefore crucial to both the evolution of life cycle traits and animal population dynamics. Changes in prey or predator abundances, the act of predation itself, and other shifts at the community level are all potential ways that climate change may impact predator-prey interactions. Interactions between predators and prey may be impacted by parasites and climate change. For instance, higher temperatures and parasite infection led to higher rates of prey intake in invading amphipods, suggesting a possible relationship between temperature and parasite infection as well as a greater ecological impact on native populations. Phenological changes can modify predator-prey interactions, just like they can in plant-herbivore interactions(Laws, 2017). Plants, herbivores, and predators frequently respond to rising temperatures by altering their phenology in positive or negative ways, which has a significant impact on the herbivores and predators. Nevertheless, compared to plant-herbivore interactions, predators at higher trophic levels are typically more negatively impacted by environmental perturbations, such as habitat modification, competition from invading species, or climate change. The interactions between all species within a community may be impacted by climate change, but as predators are essential to the structure and operation of communities, the effects of climate change on predators may have a larger influence.

It has long been known that the availability of food limits or regulates predators. The quantity of important invertebrate prey for avian predators is anticipated to be directly impacted by climate change. Food scarcity brought on by global warming may have a substantial impact on bird populations, as food availability is a limiting factor for the productivity of insectivorous birds. The relationship between asynchrony and population trends, however, may be driven by the species' sensitivity to other environmental pressures. It is becoming clearer that other factors, such as social interaction, territoriality, and interference, also play a role in the population growth of predators, rather than just the rate of prey consumption, as is typically assumed in models of predator-prey interactions. This interaction is further complicated by the differing effects of different types of predator-prey interactions and atypical dynamics. Furthermore, the studies are made more complex by the directional tendency of climate change, both in average and variance. Accurate measurements of the relationships between the temperature and cycles (prey and predator) frequently depend on limited empirical evidence. Population dynamics are determined by density, climate, and predation combined; however, it is evident that mechanistic models are necessary to fully comprehend the interactions between these elements. It seems that the combined effects of climate change and predation pressure have a deleterious effect on the population dynamics of prey species. Whether predators manage prey populations is an ongoing topic of discussion.

REFERENCES

1. Alderman, K., Turner, L. R., & Tong, S. (2012). Floods and human health: A systematic review. *Environment International*, *47*, 37–47. https://doi.org/10.1016/j.envint.2012.06.003
2. Altizer, S., Dobson, A., Hosseini, P., Hudson, P., Pascual, M., & Rohani, P. (2006). Seasonality and the dynamics of infectious diseases. *Ecology Letters*, *9*(4), 467–484. https://doi.org/10.1111/j.1461-0248.2005.00879.x
3. Antar, M., Lyu, D., Nazari, M., Shah, A., Zhou, X., & Smith, D. L. (2021). Biomass for a sustainable bioeconomy: An overview of world biomass production and utilization. *Renewable and Sustainable Energy Reviews*, *139*, 110691. https://doi.org/10.1016/j.rser.2020.110691
4. Anting, N., Md. Din, Mohd. F., Iwao, K., Ponraj, M., Jungan, K., Yong, L. Y., & Siang, A. J. L. M. (2017). Experimental evaluation of thermal performance of cool pavement material using waste tiles in tropical climate. *Energy and Buildings*, *142*, 211–219. https://doi.org/10.1016/j.enbuild.2017.03.016
5. Antoniazzi, L. R., Manzoli, D. E., Rohrmann, D., Saravia, M. J., Silvestri, L., & Beldomenico, P. M. (2011). Climate variability affects the impact of parasitic flies on Argentinean forest birds. *Journal of Zoology*, *283*(2), 126–134. https://doi.org/10.1111/j.1469-7998.2010.00753.x
6. Araujo, G. S., Silva, J. W. A. da, Cotas, J., & Pereira, L. (2022). Fish Farming Techniques: Current Situation and Trends. *Journal of Marine Science and Engineering*, *10*(11), Article 11. https://doi.org/10.3390/jmse10111598
7. Armstrong, G. A. B., & Meldrum Robertson, R. (2006). A role for octopamine in coordinating thermoprotection of an insect nervous system.

Journal of Thermal Biology, *31*(1), 149–158. https://doi.org/10.1016/j.jtherbio.2005.11.022

8. Baker, S. J. (2022). Fossil evidence that increased wildfire activity occurs in tandem with periods of global warming in Earth's past. *Earth-Science Reviews*, *224*, 103871. https://doi.org/10.1016/j.earscirev.2021.103871
9. Banwell, C., Dixon, J., Bambrick, H., Edwards, F., & Kjellström, T. (2012). Socio-cultural reflections on heat in Australia with implications for health and climate change adaptation. *Global Health Action*, *5*(1), 19277. https://doi.org/10.3402/gha.v5i0.19277
10. Barich, B. E. (2018). Eastern Borders of the Sahara and the Relations with the Nile Valley and Beyond. In *Climate Changes in the Holocene:* CRC Press.
11. Bataille, A., Levin, I. I., & Sari, E. H. R. (2018). Colonization of Parasites and Vectors. In P. G. Parker (Ed.), *Disease Ecology: Galapagos Birds and their Parasites* (pp. 45–79). Springer International Publishing. https://doi.org/10.1007/978-3-319-65909-1_3
12. Bennett, J. E., Stevens, G. A., Mathers, C. D., Bonita, R., Rehm, J., Kruk, M. E., Riley, L. M., Dain, K., Kengne, A. P., Chalkidou, K., Beagley, J., Kishore, S. P., Chen, W., Saxena, S., Bettcher, D. W., Grove, J. T., Beaglehole, R., & Ezzati, M. (2018). NCD Countdown 2030: Worldwide trends in non-communicable disease mortality and progress towards Sustainable Development Goal target 3.4. *The Lancet*, *392*(10152), 1072–1088. https://doi.org/10.1016/S0140-6736(18)31992-5
13. Bharti, A. R., Chuquiyauri, R., Brouwer, K. C., Stancil, J., LIN, J., llanos-cuentas, A., & Vinetz, J. M. (2006). Experimental infection of the neotropical malaria vector anopheles darlingi by human patient-derived plasmodium vivax in the peruvian amazon. *The American Journal of Tropical Medicine and Hygiene*, *75*(4), 610–616. https://www.ncbi.nlm.nih.gov/pmc/articles/PMC1630632/
14. Bingham, J.-P., Mitsunaga, E., & Bergeron, Z. L. (2010). Drugs from slugs—Past, present and future perspectives of ω-conotoxin research. *Chemico-Biological Interactions*, *183*(1), 1–18. https://doi.org/10.1016/j.cbi.2009.09.021
15. Blanchard-Boehm, R. D. (2004). Natural Hazards in Latin America: Tectonic Forces and Storm Fury. *The Social Studies*. https://doi.org/10.3200/TSSS.95.3.93-105
16. Bodlah, M. A., Iqbal, J., Ashiq, A., Bodlah, I., Jiang, S., Mudassir, M. A., Rasheed, M. T., & Fareen, A. G. E. (2023). Insect behavioral restraint and adaptation strategies under heat stress: An inclusive review. *Journal*

of the Saudi Society of Agricultural Sciences, *22*(6), 327–350. https://doi.org/10.1016/j.jssas.2023.02.004

17. Borna, E., Nwaru, B. I., Bjerg, A., Mincheva, R., Rådinger, M., Lundbäck, B., & Ekerljung, L. (2019). Changes in the prevalence of asthma and respiratory symptoms in western Sweden between 2008 and 2016. *Allergy*, *74*(9), 1703–1715. https://doi.org/10.1111/all.13840
18. Borzenkova, I., Zorita, E., Borisova, O., Kalniņa, L., Kisielienė, D., Koff, T., Kuznetsov, D., Lemdahl, G., Sapelko, T., Stančikaitė, M., & Subetto, D. (2015). Climate Change During the Holocene (Past 12,000 Years). In The BACC II Author Team (Ed.), *Second Assessment of Climate Change for the Baltic Sea Basin* (pp. 25–49). Springer International Publishing. https://doi.org/10.1007/978-3-319-16006-1_2
19. Boulanger, J.-P., Martinez, F., & Segura, E. C. (2006). Projection of future climate change conditions using IPCC simulations, neural networks and Bayesian statistics. Part 1: Temperature mean state and seasonal cycle in South America. *Climate Dynamics*, *27*(2), 233–259. https://doi.org/10.1007/s00382-006-0134-8
20. Bradshaw, W. E., & Holzapfel, C. M. (2017). Natural Variation and Genetics of Photoperiodism in *Wyeomyia smithii*. In S. F. Goodwin (Ed.), *Advances in Genetics* (Vol. 99, pp. 39–71). Academic Press. https://doi.org/10.1016/bs.adgen.2017.09.002
21. Brown, N. (2010). Halibut Aquaculture in North America. In *Practical Flatfish Culture and Stock Enhancement* (pp. 1–29). John Wiley & Sons, Ltd. https://doi.org/10.1002/9780813810997.ch1
22. Brunet, M., Jones, P. D., Sigró, J., Saladié, O., Aguilar, E., Moberg, A., Della□Marta, P. M., Lister, D., Walther, A., & López, D. (2007). Temporal and spatial temperature variability and change over Spain during 1850–2005. *Journal of Geophysical Research: Atmospheres*, *112*(D12), 2006JD008249. https://doi.org/10.1029/2006JD008249
23. Budd, G. M. (2008). Wet-bulb globe temperature (WBGT)—Its history and its limitations. *Journal of Science and Medicine in Sport*, *11*(1), 20–32. https://doi.org/10.1016/j.jsams.2007.07.003
24. Bush, M. B. (2002). Distributional change and conservation on the Andean flank: A palaeoecological perspective. *Global Ecology and Biogeography*, *11*(6), 463–473. https://doi.org/10.1046/j.1466-822X.2002.00305.x
25. Castagneyrol, B., Giffard, B., Valdés-Correcher, E., & Hampe, A. (2019). Tree diversity effects on leaf insect damage on pedunculate oak: The role of landscape context and forest stratum. *Forest Ecology and Management*, *433*, 287–294. https://doi.org/10.1016/j.foreco.2018.11.014

26. Charbonnier, J. (2018). Human Adaptation in Arabia: The Role of Hydraulic Technologies. In *Climate Changes in the Holocene:* CRC Press.

27. Ciota, A. T., & Kramer, L. D. (2013). Vector-Virus Interactions and Transmission Dynamics of West Nile Virus. *Viruses*, *5*(12), Article 12. https://doi.org/10.3390/v5123021

28. Collins, J. B., Jordan, B., Vidyashankar, A., Bishop, A., & Kaplan, R. M. (2022). Fenbendazole resistance in *Heterakis gallinarum*, the vector of *Histomonas meleagridis*, on a broiler breeder farm in South Carolina. *Veterinary Parasitology: Regional Studies and Reports*, *36*, 100785. https://doi.org/10.1016/j.vprsr.2022.100785

29. Crouse, C., Davidson, J., & Good, C. (2022). The effects of two water temperature regimes on Atlantic salmon (*Salmo salar*) growth performance and maturation in freshwater recirculating aquaculture systems. *Aquaculture*, *553*, 738063. https://doi.org/10.1016/j.aquaculture.2022.738063

30. Cubillo, A. M., Ferreira, J. G., Lencart-Silva, J., Taylor, N. G. H., Kennerley, A., Guilder, J., Kay, S., & Kamermans, P. (2021). Direct effects of climate change on productivity of European aquaculture. *Aquaculture International*, *29*(4), 1561–1590. https://doi.org/10.1007/s10499-021-00694-6

31. Currano, E. D., Azevedo-Schmidt, L. E., Maccracken, S. A., & Swain, A. (2021). Scars on fossil leaves: An exploration of ecological patterns in plant–insect herbivore associations during the Age of Angiosperms. *Palaeogeography, Palaeoclimatology, Palaeoecology*, *582*, 110636. https://doi.org/10.1016/j.palaeo.2021.110636

32. Cutter, S. L., Emrich, C. T., Mitchell, J. T., Boruff, B. J., Gall, M., Schmidtlein, M. C., Burton, C. G., & Melton, G. (2006). The Long Road Home: Race, Class, and Recovery from Hurricane Katrina. *Environment: Science and Policy for Sustainable Development*. https://doi.org/10.3200/ENVT.48.2.8-20

33. Dara, M., Dioguardi, M., Vazzana, M., Vazzana, I., Accardi, D., Carbonara, P., Alfonso, S., & Cammarata, M. (2022). Effects of Social Hierarchy Establishment on Stress Response and Cell Phagocytosis in Gilt-Head Sea Bream (Sparus aurata). *Fishes*, *7*(2), Article 2. https://doi.org/10.3390/fishes7020075

34. Davison, K., Dolukhanov, P., Sarson, G. R., & Shukurov, A. (2006). The role of waterways in the spread of the Neolithic. *Journal of Archaeological Science*, *33*(5), 641–652. https://doi.org/10.1016/j.jas.2005.09.017

35. Dawadi, S., & Sadof, C. S. (2023). Response of the soft scale insect

Parthenolecanium corni and its natural enemies on honeylocust trees to urban conditions. *Biological Control, 179*, 105178. https://doi.org/10.1016/j.biocontrol.2023.105178

36. Dhyani, S. K., Ram, A., Newaj, R., Handa, A. K., & Dev, I. (2020). Agroforestry for Carbon Sequestration in Tropical India. In P. K. Ghosh, S. K. Mahanta, D. Mandal, B. Mandal, & S. Ramakrishnan (Eds.), *Carbon Management in Tropical and Sub-Tropical Terrestrial Systems* (pp. 313–331). Springer. https://doi.org/10.1007/978-981-13-9628-1_19
37. Dixit, A., Carroll, S. F., & Qureshi, S. T. (2009). *Cryptococcus gattii*: An Emerging Cause of Fungal Disease in North America. *Interdisciplinary Perspectives on Infectious Diseases, 2009*, e840452. https://doi.org/10.1155/2009/840452
38. Done, A. (2012). Natural Disasters. In A. Done (Ed.), *Global Trends: Facing Up to a Changing World* (pp. 227–246). Palgrave Macmillan UK. https://doi.org/10.1057/9780230358973_13
39. Douglas, K. O., Payne, K., Sabino-Santos, G., & Agard, J. (2022). Influence of Climatic Factors on Human Hantavirus Infections in Latin America and the Caribbean: A Systematic Review. *Pathogens, 11*(1), Article 1. https://doi.org/10.3390/pathogens11010015
40. Du, Y., Zhang, Y., & Shi, J. (2019). Relationship between sea surface salinity and ocean circulation and climate change. *Science China Earth Sciences, 62*(5), 771–782. https://doi.org/10.1007/s11430-018-9276-6
41. Ebi, K. L. (2008). Adaptation costs for climate change-related cases of diarrhoeal disease, malnutrition, and malaria in 2030. *Globalization and Health, 4*(1), 9. https://doi.org/10.1186/1744-8603-4-9
42. ELLIOTT, M. (2016). Climate, ecology, and social change in prehispanic northwestern Mesoamerica. In *Settlement Ecology of the Ancient Americas*. Routledge.
43. Endo, H., & Wu, H. (2019). Biosensors for the assessment of fish health: A review. *Fisheries Science, 85*(4), 641–654. https://doi.org/10.1007/s12562-019-01318-y
44. Faurie, C., Varghese, B. M., Liu, J., & Bi, P. (2022). Association between high temperature and heatwaves with heat-related illnesses: A systematic review and meta-analysis. *Science of The Total Environment, 852*, 158332. https://doi.org/10.1016/j.scitotenv.2022.158332
45. Felsenstein, D., & Lichter, M. (2014). Social and economic vulnerability of coastal communities to sea-level rise and extreme flooding. *Natural Hazards, 71*(1), 463–491. https://doi.org/10.1007/s11069-013-0929-y

46. Fernández, D. A., Santamarina, P. E., Palazzesi, L., Tellería, M. C., & Barreda, V. D. (2021). Incursion of tropically-distributed plant taxa into high latitudes during the middle Eocene warming event: Evidence from the Río Turbio Fm, Santa Cruz, Argentina. *Review of Palaeobotany and Palynology*, *295*, 104510. https://doi.org/10.1016/j.revpalbo.2021.104510

47. Fischer Walker, C. L., Perin, J., Aryee, M. J., Boschi-Pinto, C., & Black, R. E. (2012). Diarrhea incidence in low- and middle-income countries in 1990 and 2010: A systematic review. *BMC Public Health*, *12*(1), 220. https://doi.org/10.1186/1471-2458-12-220

48. Forrest, J. R. (2016). Complex responses of insect phenology to climate change. *Current Opinion in Insect Science*, *17*, 49–54. https://doi.org/10.1016/j.cois.2016.07.002

49. Franche, C., Lindström, K., & Elmerich, C. (2009). Nitrogen-fixing bacteria associated with leguminous and non-leguminous plants. *Plant and Soil*, *321*(1), 35–59. https://doi.org/10.1007/s11104-008-9833-8

50. Gage, K. L., Burkot, T. R., Eisen, R. J., & Hayes, E. B. (2008). Climate and Vectorborne Diseases. *American Journal of Preventive Medicine*, *35*(5), 436–450. https://doi.org/10.1016/j.amepre.2008.08.030

51. Galbraith, E. D., & Eggleston, S. (2017). A lower limit to atmospheric CO2 concentrations over the past 800,000 years. *Nature Geoscience*, *10*(4), 295–298. https://doi.org/10.1038/ngeo2914

52. Gao, J., Jin, S.-S., He, Y., Luo, J.-H., Xu, C.-Q., Wu, Y.-Y., Hou, C.-S., Wang, Q., & Diao, Q.-Y. (2020). Physiological Analysis and Transcriptome Analysis of Asian Honey Bee (Apis cerana cerana) in Response to Sublethal Neonicotinoid Imidacloprid. *Insects*, *11*(11), Article 11. https://doi.org/10.3390/insects11110753

53. Gebel, H. G. K., & Wellbrock, K. (2018). Hydraulic Cultures and Hydrology under Climatic Change: North Arabian Mid-Holocene Pastoral and Proto-Oasis Land Use. In *Climate Changes in the Holocene:* CRC Press.

54. Ghude, S. D., Chate, D. M., Jena, C., Beig, G., Kumar, R., Barth, M. C., Pfister, G. G., Fadnavis, S., & Pithani, P. (2016). Premature mortality in India due to PM2.5 and ozone exposure. *Geophysical Research Letters*, *43*(9), 4650–4658. https://doi.org/10.1002/2016GL068949

55. Gibling, M. R. (2018). River Systems and the Anthropocene: A Late Pleistocene and Holocene Timeline for Human Influence. *Quaternary*, *1*(3), Article 3. https://doi.org/10.3390/quat1030021

56. Giller, K. E., Delaune, T., Silva, J. V., Descheemaeker, K., van de Ven, G., Schut, A. G. T., van Wijk, M., Hammond, J., Hochman, Z., Taulya, G.,

Chikowo, R., Narayanan, S., Kishore, A., Bresciani, F., Teixeira, H. M., Andersson, J. A., & van Ittersum, M. K. (2021). The future of farming: Who will produce our food? *Food Security*, *13*(5), 1073–1099. https://doi.org/10.1007/s12571-021-01184-6

57. Gjedrem, T., Robinson, N., & Rye, M. (2012). The importance of selective breeding in aquaculture to meet future demands for animal protein: A review. *Aquaculture*, *350–353*, 117–129. https://doi.org/10.1016/j.aquaculture.2012.04.008

58. González-Tokman, D., Córdoba-Aguilar, A., Dáttilo, W., Lira-Noriega, A., Sánchez-Guillén, R. A., & Villalobos, F. (2020). Insect responses to heat: Physiological mechanisms, evolution and ecological implications in a warming world. *Biological Reviews*, *95*(3), 802–821. https://doi.org/10.1111/brv.12588

59. Haigh, M. J., Jansky, L., & Hellin, J. (2004). Headwater deforestation: A challenge for environmental management. *Global Environmental Change*, *14*, 51–61. https://doi.org/10.1016/j.gloenvcha.2003.11.004

60. Halsey, S. J., Allan, B. F., & Miller, J. R. (2018). The role of *Ixodes scapularis*, *Borrelia burgdorferi* and wildlife hosts in Lyme disease prevalence: A quantitative review. *Ticks and Tick-Borne Diseases*, *9*(5), 1103–1114. https://doi.org/10.1016/j.ttbdis.2018.04.006

61. Harlan, S. L., Chowell, G., Yang, S., Petitti, D. B., Morales Butler, E. J., Ruddell, B. L., & Ruddell, D. M. (2014). Heat-Related Deaths in Hot Cities: Estimates of Human Tolerance to High Temperature Thresholds. *International Journal of Environmental Research and Public Health*, *11*(3), Article 3. https://doi.org/10.3390/ijerph110303304

62. Hazards (BIOHAZ), E. P. on B. (2010). Scientific Opinion on a quantitative estimation of the public health impact of setting a new target for the reduction of Salmonella in laying hens. *EFSA Journal*, *8*(4), 1546. https://doi.org/10.2903/j.efsa.2010.1546

63. Hebinck, P., & Oostindie, H. (2018). Performing food and nutritional security in Europe: Claims, promises and limitations. *Food Security*, *10*(6), 1311–1324. https://doi.org/10.1007/s12571-018-0853-9

64. Hernandez Bennetts, V., Kamarudin, K., Wiedemann, T., Kucner, T. P., Somisetty, S. L., & Lilienthal, A. J. (2019). Multi-Domain Airflow Modeling and Ventilation Characterization Using Mobile Robots, Stationary Sensors and Machine Learning. *Sensors*, *19*(5), Article 5. https://doi.org/10.3390/s19051119

65. Holmgren, E. B., & Forsgren, M. (1990). Epidemiology of Tick-borne Encephalitis in Sweden 1956–1989: A Study of 1116 Cases.

Scandinavian Journal of Infectious Diseases, *22*(3), 287–295. https://doi.org/10.3109/00365549009027050

66. Huntingford, F., Rey, S., & Quaggiotto, M.-M. (2020). Behavioural fever, fish welfare and what farmers and fishers know. *Applied Animal Behaviour Science*, *231*, 105090. https://doi.org/10.1016/j.applanim.2020.105090
67. Ioannou, L. G., Foster, J., Morris, N. B., Piil, J. F., Havenith, G., Mekjavic, I. B., Kenny, G. P., Nybo, L., & Flouris, A. D. (2022). Occupational heat strain in outdoor workers: A comprehensive review and meta-analysis. *Temperature*, *9*(1), 67–102. https://doi.org/10.1080/23328940.2022.2030634
68. Jaakkola, K., Saukkoriipi, A., Jokelainen, J., Juvonen, R., Kauppila, J., Vainio, O., Ziegler, T., Rönkkö, E., Jaakkola, J. J., Ikäheimo, T. M., & the KIAS-Study Group. (2014). Decline in temperature and humidity increases the occurrence of influenza in cold climate. *Environmental Health*, *13*(1), 22. https://doi.org/10.1186/1476-069X-13-22
69. John, T. J., Dandona, L., Sharma, V. P., & Kakkar, M. (2011). Continuing challenge of infectious diseases in India. *The Lancet*, *377*(9761), 252–269. https://doi.org/10.1016/S0140-6736(10)61265-2
70. Jump, A. S., & Peñuelas, J. (2005). Running to stand still: Adaptation and the response of plants to rapid climate change. *Ecology Letters*, *8*(9), 1010–1020. https://doi.org/10.1111/j.1461-0248.2005.00796.x
71. Junge, C. E. (1962). Global ozone budget and exchange between stratosphere and troposphere. *Tellus*, *14*(4), 363–377. https://doi.org/10.1111/j.2153-3490.1962.tb01349.x
72. Karunarathne, A., Gunnell, D., Konradsen, F., & Eddleston, M. (2020). How many premature deaths from pesticide suicide have occurred since the agricultural Green Revolution? *Clinical Toxicology*, *58*(4), 227–232. https://doi.org/10.1080/15563650.2019.1662433
73. Kass, D. C. L., Sylvester-Bradley, R., & Nygren, P. (1997). The role of nitrogen fixation and nutrient supply in some agroforestry systems of the Americas. *Soil Biology and Biochemistry*, *29*(5), 775–785. https://doi.org/10.1016/S0038-0717(96)00269-6
74. Kaygusuz, K. (2011). Energy services and energy poverty for sustainable rural development. *Renewable and Sustainable Energy Reviews*, *15*(2), 936–947. https://doi.org/10.1016/j.rser.2010.11.003
75. Kellermann, V., & van Heerwaarden, B. (2019). Terrestrial insects and climate change: Adaptive responses in key traits. *Physiological Entomology*, *44*(2), 99–115. https://doi.org/10.1111/phen.12282

76. Kennett, D. J., Lipson, M., Prufer, K. M., Mora-Marín, D., George, R. J., Rohland, N., Robinson, M., Trask, W. R., Edgar, H. H. J., Hill, E. C., Ray, E. E., Lynch, P., Moes, E., O'Donnell, L., Harper, T. K., Kate, E. J., Ramos, J., Morris, J., Gutierrez, S. M., … Reich, D. (2022). South-to-north migration preceded the advent of intensive farming in the Maya region. *Nature Communications*, *13*(1), 1530. https://doi.org/10.1038/s41467-022-29158-y
77. Kibenge, F. S. B. (2022). Chapter 1—Descriptions of major farmed aquatic animal species. In F. S. B. Kibenge, B. Baldisserotto, & R. S.-M. Chong (Eds.), *Aquaculture Pathophysiology* (pp. 1–44). Academic Press. https://doi.org/10.1016/B978-0-12-812211-2.00041-X
78. Klop, E., Omon, B., & WallisDeVries, M. F. (2015). Impact of nitrogen deposition on larval habitats: The case of the Wall Brown butterfly Lasiommata megera. *Journal of Insect Conservation*, *19*(2), 393–402. https://doi.org/10.1007/s10841-014-9748-z
79. Knell, R. J., & Thackeray, S. J. (2016). Voltinism and resilience to climate-induced phenological mismatch. *Climatic Change*, *137*(3), 525–539. https://doi.org/10.1007/s10584-016-1691-4
80. Kondratyev, K. Ya., Varotsos, C. A., & Krapivin, V. F. (Eds.). (2006). Statistics of natural disasters. In *Natural Disasters as Interactive Components of Global Ecodynamics* (pp. 1–59). Springer. https://doi.org/10.1007/3-540-37714-X_1
81. Kovats, R. S., Bouma, M. J., Hajat, S., Worrall, E., & Haines, A. (2003). El Niño and health. *The Lancet*, *362*(9394), 1481–1489. https://doi.org/10.1016/S0140-6736(03)14695-8
82. Lagadec, P. (2004). Understanding the French 2003 Heat Wave Experience: Beyond the heat, a Multi-Layered Challenge. *Journal of Contingencies and Crisis Management*, *12*(4), 160–169. https://doi.org/10.1111/j.0966-0879.2004.00446.x
83. Lancaster, L. T., Dudaniec, R. Y., Chauhan, P., Wellenreuther, M., Svensson, E. I., & Hansson, B. (2016). Gene expression under thermal stress varies across a geographical range expansion front. *Molecular Ecology*, *25*(5), 1141–1156. https://doi.org/10.1111/mec.13548
84. Laws, A. N. (2017). Climate change effects on predator–prey interactions. *Current Opinion in Insect Science*, *23*, 28–34. https://doi.org/10.1016/j.cois.2017.06.010
85. Lirette, A.-O., & Despland, E. (2021). Defensive Traits during White Spruce (Picea glauca) Leaf Ontogeny. *Insects*, *12*(7), Article 7. https://doi.org/10.3390/insects12070644

86. Liu, D., Guo, X., & Xiao, B. (2019). What causes growth of global greenhouse gas emissions? Evidence from 40 countries. *Science of The Total Environment*, *661*, 750–766. https://doi.org/10.1016/j.scitotenv.2019.01.197

87. Lopes, G., & Reis-Castro, L. (2019). A Vector in the (Re)Making: A History of Aedes aegypti as Mosquitoes that Transmit Diseases in Brazil. In C. Lynteris (Ed.), *Framing Animals as Epidemic Villains: Histories of Non-Human Disease Vectors* (pp. 147–175). Springer International Publishing. https://doi.org/10.1007/978-3-030-26795-7_6

88. Lucero, L. J., Gunn, J. D., & Scarborough, V. L. (2011). Climate Change and Classic Maya Water Management. *Water*, *3*(2), Article 2. https://doi.org/10.3390/w3020479

89. Lucero, L. J., & Larmon, J. T. (2018). Climate Change, Mesoamerica, and the Classic Maya Collapse. In *Climate Changes in the Holocene:* CRC Press.

90. Lüning, S., & Vahrenholt, F. (2019). Holocene Climate Development of North Africa and the Arabian Peninsula. In A. Bendaoud, Z. Hamimi, M. Hamoudi, S. Djemai, & B. Zoheir (Eds.), *The Geology of the Arab World—An Overview* (pp. 507–546). Springer International Publishing. https://doi.org/10.1007/978-3-319-96794-3_14

91. MacIsaac, A. (2019). *The greenhouse gas climate commitment and reversibility of peak warming from past emissions* [Masters, Concordia University]. https://spectrum.library.concordia.ca/id/eprint/985940/

92. Mafa, M. S., Castelyn, H. D., Kemp, G., & Visser, B. (2020). Delineating induced defense responses in wheat seedlings exposed to volatiles emitted by *Puccinia triticina* infected wheat. *Physiological and Molecular Plant Pathology*, *112*, 101538. https://doi.org/10.1016/j.pmpp.2020.101538

93. Maloney, S. K., & Forbes, C. F. (2011). What effect will a few degrees of climate change have on human heat balance? Implications for human activity. *International Journal of Biometeorology*, *55*(2), 147–160. https://doi.org/10.1007/s00484-010-0320-6

94. Maraseni, T. N., Deo, R. C., Qu, J., Gentle, P., & Neupane, P. R. (2018). An international comparison of rice consumption behaviours and greenhouse gas emissions from rice production. *Journal of Cleaner Production*, *172*, 2288–2300. https://doi.org/10.1016/j.jclepro.2017.11.182

95. Messner, S., Miranda, S. C., Young, E., & Hedge, N. (2011). Climate change-related impacts in the San Diego region by 2050. *Climatic Change*, *109*(1), 505–531. https://doi.org/10.1007/s10584-011-0316-1

96. Mickley, L. J. (2007). A Future Short of Breath? Possible Effects of Climate

Change on Smog. *Environment: Science and Policy for Sustainable Development*. https://doi.org/10.3200/ENVT.49.6.34-43

97. Mohajerani, A., Bakaric, J., & Jeffrey-Bailey, T. (2017). The urban heat island effect, its causes, and mitigation, with reference to the thermal properties of asphalt concrete. *Journal of Environmental Management*, *197*, 522–538. https://doi.org/10.1016/j.jenvman.2017.03.095
98. Mohanty, B., Vivekanandan, E., Mohanty, S., Mahanty, A., Trivedi, R., Tripathy, M., & Sahu, J. (2017). The Impact of Climate Change on Marine and Inland Fisheries and Aquaculture in India. In *Climate Change Impacts on Fisheries and Aquaculture* (pp. 569–601). John Wiley & Sons, Ltd. https://doi.org/10.1002/9781119154051.ch17
99. Mohino, E., Janicot, S., & Bader, J. (2011). Sahel rainfall and decadal to multi-decadal sea surface temperature variability. *Climate Dynamics*, *37*(3), 419–440. https://doi.org/10.1007/s00382-010-0867-2
100. Molina-Guzmán, L. P., Gutiérrez-Builes, L. A., & Ríos-Osorio, L. A. (2022). Models of spatial analysis for vector-borne diseases studies: A systematic review. *Veterinary World*, *15*(8), 1975–1989. https://doi.org/10.14202/vetworld.2022.1975-1989
101. Mostafa, A. N., Wheida, A., El Nazer, M., Adel, M., El Leithy, L., Siour, G., Coman, A., Borbon, A., Magdy, A. W., Omar, M., Saad-Hussein, A., & Alfaro, S. C. (2019). Past (1950–2017) and future (−2100) temperature and precipitation trends in Egypt. *Weather and Climate Extremes*, *26*, 100225. https://doi.org/10.1016/j.wace.2019.100225
102. Müller, O., & Jahn, A. (2009). Malnutrition and Maternal and Child Health. In J. Ehiri (Ed.), *Maternal and Child Health: Global Challenges, Programs, and Policies* (pp. 287–310). Springer US. https://doi.org/10.1007/b106524_16
103. Nagasundaram, M., Achyuthan, H., & Ahmad, S. M. (2014). Monsoonal changes inferred from the Middle to Late Holocene sediments off Landfall Island, North Andaman. *Arabian Journal of Geosciences*, *7*(9), 3513–3523. https://doi.org/10.1007/s12517-013-1010-6
104. Nunes, A. V., Peres, C. A., Constantino, P. de A. L., Santos, B. A., & Fischer, E. (2019). Irreplaceable socioeconomic value of wild meat extraction to local food security in rural Amazonia. *Biological Conservation*, *236*, 171–179. https://doi.org/10.1016/j.biocon.2019.05.010
105. Oh, J.-W. (2018). *Pollen Allergy in a Changing World*. Springer. https://doi.org/10.1007/978-981-10-5499-0
106. Oliver, K. M., & Higashi, C. H. V. (2021). Symbiosis in a Rapidly Changing World. In C. J. Hurst (Ed.), *Microbes: The Foundation Stone*

of the Biosphere (pp. 263–296). Springer International Publishing. https://doi.org/10.1007/978-3-030-63512-1_16

107. Oppedal, F., Dempster, T., & Stien, L. H. (2011). Environmental drivers of Atlantic salmon behaviour in sea-cages: A review. *Aquaculture*, *311*(1), 1–18. https://doi.org/10.1016/j.aquaculture.2010.11.020

108. Oster, J. L., Warken, S. F., Sekhon, N., Arienzo, M. M., & Lachniet, M. (2019). Speleothem Paleoclimatology for the Caribbean, Central America, and North America. *Quaternary*, *2*(1), Article 1. https://doi.org/10.3390/quat2010005

109. Ostro, B., Rauch, S., & Green, S. (2011). Quantifying the health impacts of future changes in temperature in California. *Environmental Research*, *111*(8), 1258–1264. https://doi.org/10.1016/j.envres.2011.08.013

110. Oyebola, O. O., & Fada, P. O. (2020). Climate change: A potential risk to cage culture and genetic structure of native fisheries in fish cage-installed inland lakes. *Lakes & Reservoirs: Science, Policy and Management for Sustainable Use*, *25*(3), 283–291. https://doi.org/10.1111/lre.12331

111. Paul, B. K. (2020). *Disaster Deaths: Trends, Causes and Determinants*. Routledge. https://doi.org/10.4324/9780429203398

112. Penning-Rowsell, E. C., Sultana, P., & Thompson, P. M. (2013). The 'last resort'? Population movement in response to climate-related hazards in Bangladesh. *Environmental Science & Policy*, *27*, S44–S59. https://doi.org/10.1016/j.envsci.2012.03.009

113. Poulin, R., & Forbes, M. R. (2012). Meta-analysis and research on host–parasite interactions: Past and future. *Evolutionary Ecology*, *26*(5), 1169–1185. https://doi.org/10.1007/s10682-011-9544-0

114. Prabhjyot-Kaur, & Hundal, S. S. (2010). Global Climate Change vis-à-vis Crop Productivity. In M. K. Jha (Ed.), *Natural and Anthropogenic Disasters: Vulnerability, Preparedness and Mitigation* (pp. 413–431). Springer Netherlands. https://doi.org/10.1007/978-90-481-2498-5_18

115. Primack, R. B., Ibáñez, I., Higuchi, H., Lee, S. D., Miller-Rushing, A. J., Wilson, A. M., & Silander, J. A. (2009). Spatial and interspecific variability in phenological responses to warming temperatures. *Biological Conservation*, *142*(11), 2569–2577. https://doi.org/10.1016/j.biocon.2009.06.003

116. Rahman, M. T., Sobur, M. A., Islam, M. S., Ievy, S., Hossain, M. J., El Zowalaty, M. E., Rahman, A. T., & Ashour, H. M. (2020). Zoonotic Diseases: Etiology, Impact, and Control. *Microorganisms*, *8*(9), Article 9. https://doi.org/10.3390/microorganisms8091405

117. Raible, C. C., Casty, C., Luterbacher, J., Pauling, A., Esper, J., Frank, D. C., Büntgen, U., Roesch, A. C., Tschuck, P., Wild, M., Vidale, P.-L., Schär, C., & Wanner, H. (2006). Climate variability—Observations, reconstructions, and model simulations for the Atlantic-European and Alpine region from 1500–2100 AD. In H. Wanner, M. Grosjean, R. Röthlisberger, & E. Xoplaki (Eds.), *Climate Variability, Predictability and Climate Risks: A European Perspective* (pp. 9–29). Springer Netherlands. https://doi.org/10.1007/978-1-4020-5714-4_2

118. Raj, A., Jhariya, M. K., Yadav, D. K., & Banerjee, A. (2020). Impact of Climate Change on Agroecosystems and Mitigation Strategies. In *Climate Change and Agroforestry Systems*. Apple Academic Press.

119. Raj, A., Jhariya, M. K., Yadav, D. K., Banerjee, A., & Meena, R. S. (2019). Agroforestry: A Holistic Approach for Agricultural Sustainability. In M. K. Jhariya, A. Banerjee, R. S. Meena, & D. K. Yadav (Eds.), *Sustainable Agriculture, Forest and Environmental Management* (pp. 101–131). Springer. https://doi.org/10.1007/978-981-13-6830-1_4

120. Rao, S. S., Singh, R. B., Takahashi, T., Juneja, L. R., Fedacko, J., & Shewale, A. R. (2019). Chapter 4—Economic Burden of Noncommunicable Diseases and Economic Cost of Functional Foods for Prevention. In R. B. Singh, R. R. Watson, & T. Takahashi (Eds.), *The Role of Functional Food Security in Global Health* (pp. 57–68). Academic Press. https://doi.org/10.1016/B978-0-12-813148-0.00004-9

121. Ravindra, K., Rattan, P., Mor, S., & Aggarwal, A. N. (2019). Generalized additive models: Building evidence of air pollution, climate change and human health. *Environment International*, *132*, 104987. https://doi.org/10.1016/j.envint.2019.104987

122. Rayne, N., & Aula, L. (2020). Livestock Manure and the Impacts on Soil Health: A Review. *Soil Systems*, *4*(4), Article 4. https://doi.org/10.3390/soilsystems4040064

123. Razafindratsima, O. H., Kamoto, J. F. M., Sills, E. O., Mutta, D. N., Song, C., Kabwe, G., Castle, S. E., Kristjanson, P. M., Ryan, C. M., Brockhaus, M., & Sunderland, T. (2021). Reviewing the evidence on the roles of forests and tree-based systems in poverty dynamics. *Forest Policy and Economics*, *131*, 102576. https://doi.org/10.1016/j.forpol.2021.102576

124. Reason, C. J. C., Hachigonta, S., & Phaladi, R. F. (2005). Interannual variability in rainy season characteristics over the Limpopo region of southern Africa. *International Journal of Climatology*, *25*(14), 1835–1853. https://doi.org/10.1002/joc.1228

125. Retchless, D., Frey, N., Wang, C., Hung, L.-S., & Yarnal, B. (2014).

Climate extremes in the United States: Recent research by physical geographers. *Physical Geography*, *35*(1), 3–21. https://doi.org/10.1080/02723646.2013.871191

126. Revel, M., Ducassou, E., Grousset, F. E., Bernasconi, S. M., Migeon, S., Revillon, S., Mascle, J., Murat, A., Zaragosi, S., & Bosch, D. (2010). 100,000 Years of African monsoon variability recorded in sediments of the Nile margin. *Quaternary Science Reviews*, *29*(11), 1342–1362. https://doi.org/10.1016/j.quascirev.2010.02.006
127. Russell, R. C. (1998). Mosquito-borne arboviruses in Australia: The current scene and implications of climate change for human health. *International Journal for Parasitology*, *28*(6), 955–969. https://doi.org/10.1016/S0020-7519(98)00053-8
128. Sarker, S., Yadav, A. K., Akter, M., Shahadat Hossain, M., Chowdhury, S. R., Kabir, Md. A., & Sharifuzzaman, S. M. (2020). Rising temperature and marine plankton community dynamics: Is warming bad? *Ecological Complexity*, *43*, 100857. https://doi.org/10.1016/j.ecocom.2020.100857
129. Sato, S. (2021). Plasmodium—A brief introduction to the parasites causing human malaria and their basic biology. *Journal of Physiological Anthropology*, *40*(1), 1. https://doi.org/10.1186/s40101-020-00251-9
130. Schilt, A., Baumgartner, M., Schwander, J., Buiron, D., Capron, E., Chappellaz, J., Loulergue, L., Schüpbach, S., Spahni, R., Fischer, H., & Stocker, T. F. (2010). Atmospheric nitrous oxide during the last 140,000 years. *Earth and Planetary Science Letters*, *300*(1), 33–43. https://doi.org/10.1016/j.epsl.2010.09.027
131. Schulze, E. D., Luyssaert, S., Ciais, P., Freibauer, A., Janssens, I. A., Soussana, J. F., Smith, P., Grace, J., Levin, I., Thiruchittampalam, B., Heimann, M., Dolman, A. J., Valentini, R., Bousquet, P., Peylin, P., Peters, W., Rödenbeck, C., Etiope, G., Vuichard, N., … Gash, J. H. (2009). Importance of methane and nitrous oxide for Europe's terrestrial greenhouse-gas balance. *Nature Geoscience*, *2*(12), 842–850. https://doi.org/10.1038/ngeo686
132. Schweiger, O., Biesmeijer, J. C., Bommarco, R., Hickler, T., Hulme, P. E., Klotz, S., Kühn, I., Moora, M., Nielsen, A., Ohlemüller, R., Petanidou, T., Potts, S. G., Pyšek, P., Stout, J. C., Sykes, M. T., Tscheulin, T., Vilà, M., Walther, G.-R., Westphal, C., … Settele, J. (2010). Multiple stressors on biotic interactions: How climate change and alien species interact to affect pollination. *Biological Reviews*, *85*(4), 777–795. https://doi.org/10.1111/j.1469-185X.2010.00125.x
133. Shackleton, D., Memon, F. A., Chen, A., Dutta, S., Kanungo, S., & Deb,

A. (2023). The changing relationship between Cholera and interannual climate variables in Kolkata over the past century. *Gut Pathogens*, *15*(1), 42. https://doi.org/10.1186/s13099-023-00565-w

134. Smoyer-Tomic, K. E., & Rainham, D. G. (2001). Beating the heat: Development and evaluation of a Canadian hot weather health-response plan. *Environmental Health Perspectives*, *109*(12), 1241–1248. https://doi.org/10.1289/ehp.011091241
135. Sorensen, J. P. R., Lapworth, D. J., Nkhuwa, D. C. W., Stuart, M. E., Gooddy, D. C., Bell, R. A., Chirwa, M., Kabika, J., Liemisa, M., Chibesa, M., & Pedley, S. (2015). Emerging contaminants in urban groundwater sources in Africa. *Water Research*, *72*, 51–63. https://doi.org/10.1016/j.watres.2014.08.002
136. Srivastava, A., Thomas, D. S. G., Durcan, J. A., & Bailey, R. M. (2020). Holocene palaeoenvironmental changes in the Thar Desert: An integrated assessment incorporating new insights from aeolian systems. *Quaternary Science Reviews*, *233*, 106214. https://doi.org/10.1016/j.quascirev.2020.106214
137. Taylor, M. W. (2014). Yellow Fever. In M. W. Taylor (Ed.), *Viruses and Man: A History of Interactions* (pp. 165–189). Springer International Publishing. https://doi.org/10.1007/978-3-319-07758-1_10
138. Tidwell, J. H. (2012). Functions and Characteristics of All Aquaculture Systems. In *Aquaculture Production Systems* (pp. 51–63). John Wiley & Sons, Ltd. https://doi.org/10.1002/9781118250105.ch3
139. Tóth, E., Gelybó, G., Dencső, M., Kása, I., Birkás, M., & Horel, Á. (2018). Soil CO2 Emissions in a Long-Term Tillage Treatment Experiment. In M. Á. Muñoz & R. Zornoza (Eds.), *Soil Management and Climate Change* (pp. 293–307). Academic Press. https://doi.org/10.1016/B978-0-12-812128-3.00019-7
140. Tsai, T. F., & Mitchell, C. J. (1988). St. Louis Encephalitis. In *The Arboviruses*. CRC Press.
141. Turner, B. L., & Klaus, H. D. (2020). *Diet, Nutrition, and Foodways on the North Coast of Peru: Bioarchaeological Perspectives on Adaptive Transitions*. Springer International Publishing. https://doi.org/10.1007/978-3-030-42614-9
142. Uprety, D. C., Reddy, V. R., & Mura, J. D. (2019). Greenhouse Gases: A Historical Perspective. In D. C. Uprety, V. R. Reddy, & J. D. Mura (Eds.), *Climate Change and Agriculture: A Historical Analysis* (pp. 31–41). Springer. https://doi.org/10.1007/978-981-13-2014-9_3

143. Valladares, F., Matesanz, S., Guilhaumon, F., Araújo, M. B., Balaguer, L., Benito-Garzón, M., Cornwell, W., Gianoli, E., van Kleunen, M., Naya, D. E., Nicotra, A. B., Poorter, H., & Zavala, M. A. (2014). The effects of phenotypic plasticity and local adaptation on forecasts of species range shifts under climate change. *Ecology Letters*, *17*(11), 1351–1364. https://doi.org/10.1111/ele.12348

144. Veizer, J., & Prokoph, A. (2015). Temperatures and oxygen isotopic composition of Phanerozoic oceans. *Earth-Science Reviews*, *146*, 92–104. https://doi.org/10.1016/j.earscirev.2015.03.008

145. Velichko, A. A., Catto, N., Drenova, A. N., Klimanov, V. A., Kremenetski, K. V., & Nechaev, V. P. (2002). Climate changes in East Europe and Siberia at the Late glacial–holocene transition. *Quaternary International*, *91*(1), 75–99. https://doi.org/10.1016/S1040-6182(01)00104-5

146. Vergé, X. P. C., De Kimpe, C., & Desjardins, R. L. (2007). Agricultural production, greenhouse gas emissions and mitigation potential. *Agricultural and Forest Meteorology*, *142*(2), 255–269. https://doi.org/10.1016/j.agrformet.2006.06.011

147. Waha, K., van Wijk, M. T., Fritz, S., See, L., Thornton, P. K., Wichern, J., & Herrero, M. (2018). Agricultural diversification as an important strategy for achieving food security in Africa. *Global Change Biology*, *24*(8), 3390–3400. https://doi.org/10.1111/gcb.14158

148. Wang, Y., Liu, X., & Herzschuh, U. (2010). Asynchronous evolution of the Indian and East Asian Summer Monsoon indicated by Holocene moisture patterns in monsoonal central Asia. *Earth-Science Reviews*, *103*(3), 135–153. https://doi.org/10.1016/j.earscirev.2010.09.004

149. Widlansky, M. E., Vita, J. A., Keyes, M. J., Larson, M. G., Hamburg, N. M., Levy, D., Mitchell, G. F., Osypiuk, E. W., Vasan, R. S., & Benjamin, E. J. (2007). Relation of Season and Temperature to Endothelium-Dependent Flow-Mediated Vasodilation in Subjects Without Clinical Evidence of Cardiovascular Disease (from the Framingham Heart Study)†. *The American Journal of Cardiology*, *100*(3), 518–523. https://doi.org/10.1016/j.amjcard.2007.03.055

150. Winckler, G., Anderson, R. F., Stute, M., & Schlosser, P. (2004). Does interplanetary dust control 100 kyr glacial cycles? *Quaternary Science Reviews*, *23*(18), 1873–1878. https://doi.org/10.1016/j.quascirev.2004.05.007

151. Worawiwat, A., Chaleeraktrakoon, C., & Sharma, A. (2021). Is increased flooding in Bangkok a result of rising local temperatures? *Journal of Hydrology X*, *13*, 100095. https://doi.org/10.1016/j.hydroa.2021.100095

152. Worku, A., Pretzsch, J., Kassa, H., & Auch, E. (2014). The significance of dry forest income for livelihood resilience: The case of the pastoralists and agro-pastoralists in the drylands of southeastern Ethiopia. *Forest Policy and Economics*, *41*, 51–59. https://doi.org/10.1016/j.forpol.2014.01.001

153. Wu, J., Porinchu, D. F., & Horn, S. P. (2019). Late Holocene hydroclimate variability in Costa Rica: Signature of the terminal classic drought and the Medieval Climate Anomaly in the northern tropical Americas. *Quaternary Science Reviews*, *215*, 144–159. https://doi.org/10.1016/j.quascirev.2019.04.023

154. Wygal, B. T., & Heidenreich, S. M. (2014). Deglaciation and Human Colonization of Northern Europe. *Journal of World Prehistory*, *27*(2), 111–144. https://doi.org/10.1007/s10963-014-9075-z

155. Yun, S., Lee, J.-W., & Yoo, J.-C. (2020). Host-parasite interaction augments climate change effect in an avian brood parasite, the lesser cuckoo *Cuculus poliocephalus*. *Global Ecology and Conservation*, *22*, e00976. https://doi.org/10.1016/j.gecco.2020.e00976

INDEX

H

I

L

M

N

O

W

Z